Water Quality Modeling

Rivers, Streams and Estuaries

NIPA® GENX ELECTRONIC RESOURCES & SOLUTIONS P. LTD.
New Delhi-110 034

Water Quality Modeling

Rivers, Streams and Estuaries

DR. R. MANIVANAN
Senior Researcher
Centre for Water and Power Research Institute
(Council of Scientific and Industrial Research)
Pune, India

NIPA® GENX ELECTRONIC RESOURCES & SOLUTIONS P. LTD.
New Delhi-110 034

NIPA® GENX ELECTRONIC
RESOURCES & SOLUTIONS P. LTD.

101,103, Vikas Surya Plaza, CU Block
L.S.C. Market, Pitam Pura, New Delhi-110 034
Ph : +91 11 27341616, 27341717, 27341718
E-mail: newindiapublishingagency@gmail.com
web: www.nipabooks.com

For customer assistance, please contact
Phone: + 91-11-27 34 17 17
Fax: + 91-11- 27 34 16 16
E-Mail: feedbacks@nipabooks.com

ISBN: 978-93-58870-67-1

Composed and Designed by NIPA®.

Dedicated to my beloved Parents and Children

PREFACE

Computer simulation modeling plays a vital role in water quality studies. Numerical simulation techniques, when successfully calibrated and properly applied, allow a numerical representation of the main ecosystem phenomena, and substantially improve our understanding of the complex interactions among different parameters such as temperature, biological oxygen demand, dissolved oxygen, salinity, and eutrophication *etc.* Simulation models are capable of providing an objective evaluation of the consequences of future growth and pollution control measures. Therefore, simulation models are essential tools for water quality studies for urban, rural and industrial planning and decision making.

Water pollution issues represent a growing public concern and a major topic for scientific and industrial research. Water pollution, a problem that a few decades ago used to affect only the most industrial countries in the world, has expended to virtually all countries. Recently India also witnessed growing demand for good quality water for best designated uses.

Water is an important element for life on the earth. It is an essential natural resource for environmental sustenance, agricultural productivity, industrial growth, power production and enrichment and renewal of land and air. For easy assess to the water, the first human settlement occurred along the banks of rivers and ensuing civilizations flourished in river valleys. For example, Indian civilization of Mohanjodaro and Harappa developed in the Indus valley, the Egyptian civilization in the Nile valley, the Chinese civilization in the Yangtze valley, the Babylonian civilization between the Tigris and the Euphrates rivers, Aztec Maya and Incas civilizations in the basin of Amazon River.

Supply of water at any given location is limited and the limits vary considerably. The unprecedented economic development, social transformation, industrial flourishment and technological improvement that have occurred in the world have been in part due to human ingenuity in providing water to various designated best uses, often through well-scientific and engineered strategies for usage, storing of water and transporting it. There is a direct correlation between the development of these strategies and national development, as exemplified by the grouping of nations of the world into developed countries for example USA, Britain, Japan etc., developing countries like India, Sri Lanka, Pakistan etc., and under developed countries such as Kenya, Lozetho etc. From the middle of the nineteenth century to the beginning of twenty first century, hundreds of water resources projects, water harvesting projects, water pollution projects, water recycling projects, river linking projects, desalination projects for fresh water from ocean and water quality modeling studies were developed and built throughout the world including India, primarily for the purposes to predict how extent of pollution affected in the water resources such as rivers, lakes, reservoirs, coastal as well as oceans in the world. In India, water quality modeling studies were done for ecosystems from fresh water to marine water ecosystems for example Tehri reservoir, Chilka lake, Dal lake at Kashmir, Kodaikanal lake, Ooty lake, rivers like Ganges, Narmada, Kaveri, and all coastal regions like Chennai, Visakhapatanam, Hoogly estuary, Mangalore coast, Kongan coast, Gujarat coast. These projects were entirely funded or heavily subsidized by national and international agencies in recognition of the crucial role that water plays in an important role in the human life.

The book contains eighteen chapters reflecting objectives on various aspects of water quality modeling for rivers, streams, reservoirs, lakes, impoundments, coastal regions, seas and oceans. The chapter one explains about the importance of water, sources of water pollution, some organic and inorganic water pollutants, water quality meaning and scope, types of modeling, calibration methods, qualitative and quantitative models. The chapter two depicts modeling basics. It includes historical development, types of water quality models, model classification different water ecosystems, factors governing water quality models. The chapter three deal with processes and development. It includes model architecture, equation, chemical, physical and biological processes in the water ecosystem, computer programming, boundary and initial conditions, calibration and validation. Chapter four includes modeling application to different water bodies, spatial discretisation, concentration modeling and CORMIX. The chapter five explains methodologies. Chapter six deal with modeling of rivers and streams. Coastal and estuaries modeling will be provided in chapter seven. Eighth will be the lakes and impoundments modeling. Ocean and seas are provided with chapter nine and followed by water quality modeling of biological oxygen demand and dissolved oxygen. Temperature modeling is allocated in the chapter eleven. The chapter twelve includes nitrogen and phosphorus modeling. Eutrophication modeling explained in chapter thirteen. Oil spill and Tsunami modeling chapters are provided in fourteen and fifteen respectively. Chapter sixteen depicts about littoral drift modeling. An ecohydrological modeling and commercially available models are included in chapter seventeen and eighteen. A special technical terms and definitions are given as appendix.

A common feature of all the chapters is the emphasis given to the application of new computer technologies, such as super computers and graphical workstations and the use of advanced software tools for real time user friendly interactions between the models and the users in the subject of water quality models.

Many Engineers and Scientists are good enough to send me their research publication and their contribution to this work is highly acknowledged. A good deal of tiresome labour involved in putting togather the MS was shared by many friends, too numerous to be named individually, and I thank each one of them for cheerfully giving me their valuable time and knowledge. To my wife Mrs. S. Mekalamanivanan who helped in many ways, I wish to record my greatful thanks to her.

This book will be the interest of research scientists, NGOs, hydraulic engineer, environmental activists, ecologists, industrialists, consultants, research scholars and practitioners in the field of environmental hydraulics and engineering. It will also be interest of management of coastal hydrodynamics and coastal processes. This book will be useful for Graduate and Postgraduate students of environmental sciences, oceanography, marine sciences, natural resources as well as those who are willing to conduct further research in environmental hydraulics and engineering, hydrology and water resource planning, development and management.

1 May, 2008 **DR. R. MANIVANAN**
Pune

CONTENTS

Chapter 1

Introduction

A model is something that mimics the behavior of some real system of interest. Our interest here will be in water quality models, which are usually equations, and which predict some result based on factors that influence that result. A water quality model is a representation of a real system. The essence of a good water quality model is that it is simple in design and exhibits the basic properties of the real system that we are attempting to understand. The model should be testable against empirical data. The comparisons of the model to the real system should ideally lead to improved mathematical models. The model may suggest improved experiments to highlight a particular aspect of the problem, which in turn may improve the collection of data. Thus, modeling itself is an evolutionary process, which continues toward learning more about certain processes rather than finding an absolute reality.

Water pollution is a large set of adverse effects upon water bodies such as lakes, rivers, oceans, groundwater caused by human activities. Although natural phenomena such as volcanoes, storms, earthquakes *etc.* also cause major changes in water quality and the ecological status of water while these are not deemed to be pollution. Water pollution has many causes and characteristics. Increases in nutrient loading may lead to eutrophication. Organic wastes such as domestic sewage and farm waste impose high oxygen demands on the receiving water leading to oxygen depletion with potentially severe impacts on the whole ecosystem. Industrial discharge a variety of pollutants in their wastewater including heavy metals, organic toxins, oils, nutrients, and solids. Discharges can also have thermal effects, especially those from thermal and nuclear power stations, and these too reduce the available oxygen. Siltbearing runoff from many activities including construction sites, forestry and farms can inhibit the penetration of sunlight through the water column restricting photosynthesis and causing blanketing of the lake or river bed which in turns damages the ecosystems the process is called as eutrophication.

Pollutants in water include a wide spectrum of chemicals, pathogens, and physical chemistry or sensory changes. Many of the chemical substances are toxic or even carcinogenic. Pathogens can obviously produce waterborne diseases in either human or animal hosts. Alteration of water's physical chemistry include acidity, conductivity, temperature, and excessive nutrient loading. Even many of the municipal water supplies in developing countries can present health risks leads to many diseases.

1.1 Sources of Water Pollution

Principal sources of water pollution are:

- industrial discharge of chemical wastes and byproducts
- discharge of poorly treated or untreated sewage
- surface runoff containing pesticides
- slash and burn farming practice, which is often an element within shifting cultivation agricultural systems
- surface runoff containing spilled petroleum products
- surface runoff from construction sites, farms, or paved and other impervious surfaces *e.g.* silt
- discharge of contaminated and/or heated water used for nuclear and thermal power stations
- acid rain caused by industrial discharge of sulfur dioxide and nitrous oxide
- excess nutrients added by runoff containing detergents and fertilizers, pesticides
- underground storage tank leakage, leading to soil contamination, hence aquifer contamination

1.2 Contaminants

Contaminants are classified into organic and inorganic substances based on their constituents

1.2.1 Organic water pollutants

- insecticides and herbicides, a huge range of organohalide and other chemicals
- bacteria, often is from sewage or livestock operations
- food processing waste, including pathogens, organic waste
- tree and brush debris from logging operations
- VOCs (Volatile Organic Compounds, industrial solvents) from improper storage

1.2.2 Inorganic water pollutants

- heavy metals including acid mine drainage, tanning industries
- acidity caused by industrial discharges
- chemical waste as industrial by-products
- fertilizers, in runoff from agriculture including nitrates and phosphates
- silt in surface runoff from construction sites, logging, slash and burn practices or land clearing sites

Most water pollutants are eventually carried by the rivers into the oceans. In some areas of the India the influence can be traced hundred miles from the mouth by studies using hydrology transport models. Advanced computer models such as SWMM, MIKE21 HD, MIKE21 AD, PA, ECOLAB, ME or the DSSAM Model have been used in many locations worldwide to examine the fate of pollutants in aquatic systems. Indicator filter feeding species such as copepods have also been used to study pollutant fates. The highest toxin loads are not directly at the mouth of a river, but 100 kilometers distance, since several days are required for incorporation into planktonic tissue. The river discharge flows further distance along the coast due to coriolis force. Further distance then are areas of oxygen depletion, caused by chemicals using up oxygen and by algae blooms, caused by excess nutrients from algal cell death and decomposition. Fish and shellfish kills have been studied, because toxins climb the foodchain after small fish consume copepods, then large fish eat smaller fish, etc. Each step up the food chain concentrates certain toxins like heavy metals and DDT by approximately a factor of ten.

Many chemicals undergo reactive decay or change especially over long periods of time in groundwater reservoirs. A noteworthy class of such chemicals are the chlorinated hydrocarbons such as trichloroethylene and tetrachloroethylene used in the dry cleaning industry. Both of these chemicals, which are carcinogens themselves, undergo partial decomposition reactions leading to new hazardous chemicals.

Groundwater pollution is much more difficult to abate than surface pollution because groundwater can move great distances through unseen aquifers. Nonporous aquifers such as clays partially purify water of bacteria by simple filtration, dilution, and, in some cases, chemical reactions and biological activity: however, in some cases, the pollutants merely transform to soil contaminants. Groundwater that moves through cracks and caverns is not filtered and can be transported as easily as surface water.

There are a variety of secondary effects stemming not from the original pollutant, but a derivative condition. Some of these secondary impacts are:

- Silt bearing surface runoff from downhill area can inhibit the penetration of sunlight through the water column, hampering photosynthesis in aquatic plants.
- Thermal pollution can induce fish kills and invasion by new thermophyllic species

Water quality modelling studies defines the prediction of water pollution in the water ecosystem using mathematical modeling techniques. A typical water quality model consists of collection of formulations representing physical, chemical, biological and microbiological machanisms that determine fate and transport of pollutants in a water body. Models are available for individual components of the water quality such as BOD, DO,COD, Nutrients, Temperature; there also exist types of ecosystems like river, lake, stream, pond, coastal, ocean models addressing transport of pollutants in different ecosystems. Thus, water quality modeling involves the prediction of water pollution using mathematical simulation techniques. A typical water quality model consists of a collection of formulations representing physical machanisms that determine fate and transport of pollutants in a water body.

1.3 Water Quality Parameters

It is the chemical and physical characterization of water. The primary bases for such characterization are parameters which relate to potability, safety of human contact and for health of ecosystems. The vast majority of surface water on the planet is neither potable nor toxic. This remains true even if sea water in the oceans isn't counted. Another general perception of water quality is that of a simple property that tells whether water is polluted or not. Infact, water quality is a very complex subject, in part because water is a complex medium intrinsically tied to the ecology of the planet.

Contaminants that may be in untreated water include microorganisms such as viruses and bacteria; inorganic contaminants such as salts and metals; pesticides and herbicides; organic chemical contaminants from industrial processes and petroleum use; and radioactive contaminants. Water quality depends on the local geology and ecosystem, as well as human uses such as sewage dispersion, industrial pollution, use of bodies water as a heat sink, and overuse.

The Environmental Protection Agency prescribes regulations that limit the amount of certain contaminants in the water provided by public water systems for tap water. Food and Drug Administration regulations establish limits for contaminants in bottled water that must provide the same protection for public health. Drinking water, including bottled water, may reasonably be expected to contain at least small amounts of some contaminants. The presence of these contaminants does not necessarily indicate that the water poses a health risk.

Some people use water purification technology to remove contaminants from the municipal water supply they get in their homes, or from local pumps or bodies of water. For people who get water from a local stream, lake, or groundwater, their drinking water is not filtered by the local government, or community.

Toxic substances and high populations of certain microorganisms can present a health hazard for non-drinking purposes such as irrigation, swimming, fishing, rafting, boating, and industrial uses. These conditions may also impact wildlife which use the water for drinking or as a habitat.

Interest by individuals and volunteer groups in making local water quality observations is high, and an understanding of the basic chemistry of many water quality parameters is an essential first step in making good measurements and can effect the model to a great extent (Table 1.1). Most citizens harbor great concern over the purity of their drinking water, but there is far more to water quality than water treatment for human consumption.

Statements to the effect that "uses must be preserved" are included within water quality regulations because they provide for broad interpretation of water quality results, while preserving the ultimate goal of the regulations. Technical measures of water quality, that is, the values obtained when making water quality measurements are always subject to interpretation from multiple perspectives. Is it reasonable to expect a river to be pristine in a landscape that no longer is? If a river has always carried sediment, is it polluted even if the cause is not man induced? Can water quality be maintained when water quantity cannot? The questions that arise from consideration of water quality relative to human uses of the water become more complex when consideration must also be given to conditions required to sustain aquatic biota. Yet inherent in the concept of preserving uses is a mandate that waterways must be much more than conduits for a fluid we might want to drink, fill our swimming pool with, or carry our wastes out of town.

The complexity of water quality as a subject is reflected in the many types of measurements of water and wastewater quality. These measurements include:

- Conductivity
- Dissolved Oxygen (DO)
- pH
- Color of water
- Taste and odor
- Turbidity
- Total Suspended solids (TSS)
- Chemical Oxygen Demand (COD)
- Biochemical Oxygen Demand (BOD)

- Microorganisms such as fecal coliform bacteria (Escherichia coli), Cryptosporidium, and Giardia lamblia
- Nutrients (fertilizers: nitrates, phosphates)
- Dissolved metals and metalloids (lead, mercury, arsenic, etc.)
- Dissolved organics
- Temperature
- Pesticides
- Heavy Metals

Table 1.1: Water Quality Parameters and Their Effect

Parameter	Effect on Water body
Dissolved oxygen	Usually decreases as discharge increases. Used as a water quality indicator in most
Biochemical oxygen demand (BOD)	A measure of oxygen-reducing potential for waterborne discharges. Used in most water quality models.
Temperature	Often increased by discharges, especially from electric power plants. Relatively easy to model.
Ammonia nitrogen	Reduces dissolved oxygen concentrations and adds nitrate to water. Can be predicted by most water quality models.
Algal concentration	Increases with pollution, especially nitrates and phosphates. Predicted by moderately complex models.
Coliform bacteria	An indicator of contamination from sewage and animal waste
Nitrates	A nutrient for algal growth and a health hazard at very high concentrations in drinking water. Predicted by moderately complex models.
Phosphates	Nutrient for algal growth. Predicted by moderately complex models.
Toxic organic Compounds	A wide variety of organic (carbon-based) compounds can affect aquatic life and may be directly hazardous to humans. Usually very difficult to model.
Heavy metals	Substances containing lead, mercury, cadmium, and other metals can cause both ecological and human health problems. Difficult to model in detail.

Most of the simple measurements listed above can be made on-site, in direct contact with the water source in question. More complex measurements that must be made in a lab setting require a water sample to be collected, preserved, and analyzed at another location. Making these complex measurements can be expensive, and it is important to understand in advance the reasons for making such complex measurements.

1.4 Modeling-Meaning and Scope

A water quality model is a abstract model that uses mathematical language to describe the behaviour of a system. Water quality models are used particularly in the engineering sciences like civil, electronic, environmental sciences and engineering disciplines (such as physics, biology, and electrical engineering) but also in the social sciences (economics, sociology and political science); physicists, engineers, computer scientists, and economists use mathematical models most extensively. It defined a mathematical model as a representation of the essential aspects of an existing system which presents knowledge of that system in usable form.

1.5 Background

Often when scientist analyze a system to be controlled or optimized, they use a mathematical model. In analysis, engineers/ scientists can build a descriptive model of the system as a hypothesis of how the system could work, or try to estimate how an unforeseeable event could affect the system. Similarly, in control of a system, engineers/scientists can try out different control approaches in simulations.

A mathematical model usually describes a system by a set of variables and a set of equations that establish relationships between the variables. The values of the variables can be practically anything; real or integer numbers, boolean values or strings, for example. The variables represent some properties of the system, for example, measured system outputs often in the form of signals, timing data, counters, event occurrence. The actual model is the set of functions that describe the relations between the different variables.

There are six basic groups of variables: decision variables, input variables, state variables, exogenous variables, random variables, and output variables. Since there can be many variables of each type, the variables are generally represented by vectors.

Decision variables are sometimes known as independent variables. Exogenous variables are sometimes known as parameters or constants. The variables are not independent of each other as the state variables are dependent on the decision, input, random, and exogenous variables. Furthermore, the output variables are dependent on the state of the system.

Objectives and constraints of the system and its users can be represented as functions of the output variables or state variables. The objective functions will depend on the perspective of the model's user. Depending on the context, an objective function is also known as an index of performance, as it is some measure of interest to the user. Although there is no limit to the number of objective functions and constraints a model can have, using or optimizing the model becomes more involved.

1.6 Interpretation of Modeling Results

The results of water quality modeling studies must be interpreted from a professional perspective. The following approaches can provide a basis for such interpretations; they should be used collectively and not singularly:

(1) Existing water quality standards or criteria either in country BIS or WHO or UNEP standards or criteria

(2) Case study comparisons which are focused on experienced consequences of other water quality modeling projects

(3) Changes in parameters or characteristics when compared to baseline conditions; need to consider natural variability of parameters and characteristics; could calculate percentage changes from baseline conditions; and

(4) Aquatic ecological consequences of water quality changes; examples include implications for algae, rooted or floating plants, benthic organisms, and fish. (Fig.1.1)

1.7 Monitoring as a Supplement to Modeling

In order to water quality modeling efforts to be successful, water quality monitoring should be considered as a required supplement.

As noted earlier, monitoring is needed in order to calibrate/verify the modeling efforts at various times in the life cycle of a project. Planning a water quality monitoring program can be complicated due to the large number of items to consider.

Examples of items which must be addressed during the planning and operation of a monitoring program include:

(1) Delineation of monitoring objectives.

(2) Selection of water quality parameters to be monitored (including the possible use of algae bioassays).

(3) Monitoring station location and depth of sampling.

(4) Frequency of sampling.

(5) Operational consideration for monitoring (including sample storage, field testing, laboratory analytical methods, and data storage and retrieval).

(6) Preparation of monitoring reports and the usage of the information.

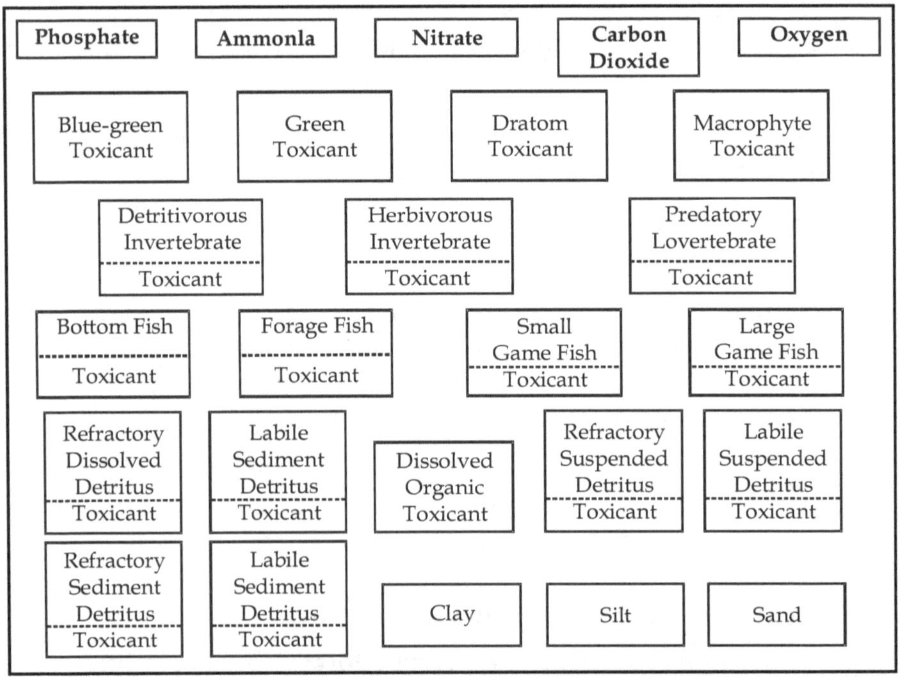

Fig. 1.1 Variables Involved in the Water Quality Modeling

1.8 Other Related Issues

In addition to direct water quality changes within water ecosystems used for hydropower, water supply, and/or flood control, there are other related issues associated with general water resources concerns. These related issues include health impacts, changes in reservoir biology, alterations in the local microclimate, and downstream effects.

(1) Health impact concerns are a result of construction work influx, people relocation, and the reservoir serving as a breeding place for mosquitoes and other disease factors. Water related diseases can be classified into waterborne diseases, waterbased diseases, water related diseases, and water washed diseases.

(2) Changes in reservoir biology as a result of water quality changes; these biological changes can occur on phytoplankton, aquatic plants, benthic organisms, and fish.

(3) Microclimate effects of the reservoir; these effects are largely dependent upon the size of the reservoir.

(4) Downstream effects of the reservoir on water quality and stream biology; downstream effects can also occur on estuaries as a result of flow decreases, water quality changes, and stream biology changes.

1.9 Types of Ecosystems

The water ecosystems are classified into various types based on their flow, salty nature, volume etc. the following chapters are discussed about the nature and modeling.

1.9.1 Modeling - A river ecosystem

A river is a large natural waterway. The source of a river may be a lake, a spring, or a collection of small streams, known as headwaters. From their source, all rivers flow downhill, typically terminating in the ocean. The mouth, or lower end, of a river is known as its base level (Fig. 1.2).

A river's water is confined to a channel, made up of a stream bed between banks. Most rainfall on land passes through a river on its way to the ocean. Smaller side streams that join a river are tributaries.

The scientific term for any flowing natural waterway is a stream; so in technical language, the term river is just a shorthand way to refer to a large stream. Rivers throughout history have been the source and support for civilization, and many major cities today are near a river of some sort.

Fig. 1.2 A Typical River Ecosystem

1.9.2 Modeling- A coastal ecosystem

The coast is one of the most important boundaries in the world. It separates the marine part of the world from the continents. It separates the salt-water communities from fresh water communities; it separates land life from water borne life but amphibians living both environments. Some of the marine organisms move towards estuary for reproduction with less water salinity. The coastline as such is not a stable. It fluctuates in place over considerable distances due to erosion and sedimentation, but also due to the relative change of sea level with respect to land. The coastal zone is thus a very mobile part of the earth from a geographical point of view. Due to its specific geophysical properties, the coast is also a very important area for the development of flora and fauna. There is hardly any other part of the world that houses such a variety of species, both in flora and fauna. The abundance of opportunities along the coast has also attracted a lot of human activities. The coastal zone has become an indispensable resource for mankind. It provides excellent opportunities for fisheries, housing, recreation, transport, defense purpose, industrial activities and many others. The influx of man with their social and economic activities, their needs for safe housing, food, drinking water, and with the tremendous quantities of waste they are producing poses at the same time a serious threat to the unique features of the coastal zone. It is therefore necessary to study this valuable region so as to prevent that competing activities are gradually causing depletion of the resources in the coastline. The following figures (Fig. 1.3) show the typical coastal area of rocky coast and sandy coast.

Fig. 1.3 Typical Coastal Area of Rocky and Sandy Coasts

1.9.3 Modeling – Lakes and impoundments

A lake is a body of water or other liquid of considerable size surrounded entirely by land. A vast majority of lakes on earth are of fresh water, and most of them lie in the Northern Hemisphere at higher latitudes. In ecology, the environment of a lake is referred to as lacustrine. Large lakes are occasionally referred to as "inland seas" and small seas are occasionally referred to as lakes. Smaller lakes tend to put the word "lake" after the name, as in Green Lake, while larger lakes often invert the word order. The term lake is also used to describe a feature such as Lake Eyre, which is a dry basin most of the time but may become filled under seasonal conditions of heavy rainfall. Many lakes are artificial and are constructed for hydroelectric power supply, recreational purposes, industrial use, agricultural use, or domestic water supply (Fig. 1.4).

Fig. 1.4. A Typical Lake Ecosystem

1.9.4 Modeling - Seas and oceans

Oceans (from Okeanos in Greek) are saline waters that cover almost three quarters (71%) of the surface of the earth.

Fig. 1.5. A Typical Oceans and Seas

The area of the oceans is 361 million sq. km., and nearly half of the world's marine waters are over 3,000 meters (9,800 ft) deep. Though somewhat arbitrarily divided into several "separate" oceans, these oceans are in fact one global, interconnected body of salt water, often called the World Ocean. Fig.1.5 shows the typical ocean area. The major divisions are defined in part by the continents and a variety of archipelagos, and are labeled the Pacific Ocean, the Atlantic Ocean, the Indian Ocean, the Southern Ocean, and the Arctic Ocean. Smaller regions of the oceans are called seas, gulfs, straits and other names. The following chapters will be discussed about the modeling of water quality parameters with their ecosystems, for example river water quality modeling, lake water quality modleing, ocean modleing.

❑❑❑

Chapter 2

Water Quality Modeling

Basics

2.1 Historical Development

We have seen in Chapter 1 that modeling is about establishing cause effect relationship. Perhaps one of the earliest attempts in this direction is made as early as 1848 in Germany where it was tried to establish a relation between purity and pollution to the occurrence of certain biological organism. But as far as water quality modeling is concerned the first major breakthrough in this direction is the effort of Streetor and Phelps (1925) in oxygen sag curve. They developed model for estimating DO and BOD in Ohio River. The work of O'Conner (1962) for bacteria models are major milestone in pre-computer era modeling.

With the advent of computer during 1960s complicated kinetics and time variable simulation came into being. Most of the workers presented their efforts in terms of linear programming based water management models concerning waste discharge. Thomann(1963) has formulated linear programming based models for dissolved oxygen in an estuary. So a stage is set for study of eutrophication through modeling, but the problem is determination of values of various parameters in these equations.

During 1970s models expanded in terms of complexities mainly because of enhanced computational facilities. Two dimensional and dynamic framework has started replacing one dimensional steady state configurations. In the same year, Di Toro presented an analysis of the steady state two dimensional concentration distribution that results in the horizontal plane from the continuous discharge of a non-conservative substances (DO and BOD) into an infinite body of water.

2.2 Types of Mathematical Models

There are numerous water quality models that exhibit a wide range of technical sophistication and input data requirements (Canter, 1985). Three key points are relevant in the selection and usage of these numerical models: (i) simple models should be used in the case where water quality and flow data is minimal; (ii) qualitative models can be useful for planning monitoring programs and anticipating impacts in the absence of site specific data; and (iii) sophisticated multi-dimensional models should only be used when the available data is sufficient to justify the usage of such models. Accordingly, consideration should be given to the selection of appropriate models for each of the temporal phases mentioned earlier. Four types of models to be briefly summarized herein include physical models, qualitative models, simple quantitative models; and quantitative models.

2.3 Physical Models

Physical models can be built and used to enable the physical measurement of water quality changes. Examples include:

(1) Scale models of segments of a reservoir 10 meter diameter and 10 meter deep tanks are an example can use to determine important processes and develop rate information.

(2) Laboratory microcosms to focus on understanding specific processes and developing rate information 0.5 to 1 meter cubes or rectangles to simulate water quality of a water body.

2.4 Qualitative Models

Qualitative predictions of water quality changes in a proposed water body can be based on using data on experienced changes from project of site. It would be best to look at common patterns of change from several studies. Expert judgment must be used to qualitatively infer changes in water quality; using several pertinent professionals can best do this, examples include environmental engineers, chemists, and aquatic biologists.

2.5 Quantitative Models

There are several simple models and relationships, which can be used to estimate water quality changes in reservoirs. Examples include:

(1) Calculate mass balances of chemical inputs and outputs to a water body; based upon calculations related to water body size (volume) and flow rate, and assumptions related to in water body processes, average in water body concentrations of specific water quality constituents can be calculated; an example of the major nutrient inputs and exports for a water body include (Interim Mekong Committee, 1982):

 (a) inputs from five sources watershed surface drainage, ground water inflow, reservoir sediments, atmospheric precipitation and dry fallout, and direct lakeside introductions;

 (b) exports from water phase via three routes movement through water body and to downstream, sedimentation, and harvesting of biomass.

(2) Can express pollutant loading from various point and nonpoint sources via calculation of population equivalent (Canter, 1977).

(3) Utilize densimetric Froude number to anticipate the propensity of water body to thermally stratify (Canter, 1985).

(4) Can use simple mixing model for entire water body or for segments of water body, in this case uniform mixing of all constituents is assumed to occur, could then account for various reservoir processes described earlier.

(5) Based on rational assumptions related to reaeration in a reservoir versus a flowing stream, could calculate the change in waste assimilative capacity by assuming certain organic matter decomposition rates (Canter, 1977).

(6) Can account for bacterial die-away by using Chick's law and adjusting the die away rate for reservoir conditions (Canter, 1977).

Quantitative models typically involve the consideration of multiple influencing factors and processes in a reservoir. Quantitative models also typically consider reservoir dynamics in lieu of steady state conditions; models for transient conditions can be for continuous time or discrete time. To serve as an example, the oxygen relationships in a reservoir can be modeled by considering decomposition, reaeration, benthic oxygen demand, and photosynthesis.

Quantitative models can be one-dimensional (x direction), two-dimensional (x and y directions), or three-dimensional (x, y and z directions); the multi-dimensional models are more complicated from a mathematical perspective. In addition, they are more data intensive.

2.6 Model Classifications

Mathematical models can be classified in several ways, some of which are described below.

1. *Linear vs. nonlinear:* Mathematical models are usually composed by variables, which are abstractions of quantities of interest in the described systems, and operators that act on these variables, which can be algebraic operators, functions, differential operators, etc. If all the operators in a mathematical model present linearity, the resulting mathematical model is defined as linear. A model is considered to be nonlinear otherwise. In a mathematical programming model, if the objective functions and constraints are represented entirely by linear equations, then the model is regarded as a linear model. If one or more of the objective functions or constraints are represented with a nonlinear equation, then the model is known as a nonlinear model.
2. *Deterministic vs. probabilistic (stochastic):* A deterministic model is one in which every set of variable states is uniquely determined by parameters in the model and by sets of previous states of these variables. Therefore, deterministic models perform the same way for a given set of initial conditions. Conversely, in a stochastic model, randomness is present, and variable states are not described by unique values, but rather by probability distributions.
3. *Mechanistic vs emperical:* Mathematical models can be mechanistic or empirical. A mechanistic model is one that is derived from more fundamental principles. For example, we could use the laws of physics to derive a mechanistic model that predicts the motion of a building in an earthquake. In many cases, a system is too complicated to predict from fundamental principles. Almost anything to do with human behavior, such as economic systems, will fall into this category. But many physical systems can also be beyond mechanistic modeling. Some aspects of weather prediction would be an example.

4. *Static vs. dynamic:* A static model does not account for the element of time, while a dynamic model does. Dynamic models typically are represented with difference equations or differential equations.

5. *Lumped parameters vs. distributed parameters:* If the model is homogeneous (consistent state throughout the entire system) the parameters are lumped. If the model is heterogeneous (varying state within the system), then the parameters are distributed. Distributed parameters are typically represented with partial differential equations.

6. *Discrete vs. Continuous Event Models:* Another label used to describe simulation models is discrete and continuous. Discrete models refer to simulation models that have distinct increments for time, such as years, months, weeks or days. Continuous models have no time steps or there are so many periods in the time horizon that periods are not differentiated.

7. *Black box vs white box:* Mathematical modelling problems are often classified into black box or white box models, according to how much a priori information is available of the system. A black-box model is a system of which there is no a priori information available. A white box model (also called glass box or clear box) is a system where all necessary information is available. Practically all systems are somewhere between the black-box and white-box models, so this concept only works as an intuitive guide for approach.

Usually it is preferable to use as much a priori information as possible to make the model more accurate. Therefore the white box models are usually considered easier, because if you have used the information correctly, then the model will behave correctly. Often the a priori information comes in forms of knowing the type of functions relating different variables. For example, if we make a model of how a medicine works in a human system, we know that usually the amount of medicine in the blood is an exponentially decaying function. But we are still left with several unknown parameters; how rapidly does the medicine amount decay, and what is the initial amount of medicine in blood? This example is therefore not a completely white box model. These parameters have to be estimated through some means before one can use the model.

In black box models one tries to estimate both the functional form of relations between variables and the numerical parameters in those functions. Using a priori information we could end up, for example, with a set of functions that probably could describe the system adequately. If there is no a priori information we would try to use functions as general as possible to cover all different models. An often used approach for black-box models are neural networks which usually do not assume almost anything about the incoming data. The problem with using a large set of functions to describe a system is that estimating the parameters becomes increasingly difficult when the amount of parameters increases.

2.7 Complexity

Another basic issue is the complexity of a model. If we are, for example, modelling the flight of an aircraft, we can embed each mechanical part of the aircraft into our model and would thus acquire an almost white box model of the system. However, the computational cost of adding such a huge amount of detail will effectively inhibit the usage of such a model. Additionally, the uncertainty will increase due to an overly complex system, because each separate part induces some amount of variance into the model. It is therefore usually appropriate to make some approximations to reduce the model to a sensible size. Engineers often can accept some approximations in order to get a more robust and simple model. For example Newton's classical mechanics is an approximated model of the real world. Still, Newton's model is quite sufficient for most ordinary-life situations, that is, as long as particle speeds are well below the speed of light, and we study macro particles only.

2.8 Training

Any model which is not pure white box contains some parameters that can be used to fit the model to the system it shall describe. If the modeling is done by a neural network, the optimization of parameters is called training. In more conventional modeling through explicitly given mathematical functions, parameters are determined by curve fitting.

2.9 Model Evaluation

An important part of the modelling process is the evaluation of an acquired model. How do we know if a mathematical model describes the system well? This is not an easy question to answer. Usually the engineer or scientist has a set of measurements from the system which are used in creating the model. Then, if the model is built well, the model will adequately show the relations between system variables for the measurements at hand. The question then becomes: How do we know that the measurement data are a representative set of possible values? Does the model describe well the properties of the system between the measurement data? Does the model describe well events outside the measurement data?

A common approach is to split the measured data into two parts; training data and verification data. The training data are used to train the model, that is, to estimate the model parameters. The verification data are used to evaluate model performance. Assuming that the training data and verification data are not the same, we can assume that if the model describes the verification data well, then the model describes the real system well.

However, this still leaves the extrapolation question open. How well does this model describe events outside the measured data? Consider again Newtonian classical mechanics model. Newton made his measurements without advanced equipment, so he could not measure properties of particles travelling at speeds close to the speed of light. Likewise, he did not measure the movements of molecules and other small particles, but macro particles only. It is then not surprising that his model does not extrapolate well into these domains, even though his model is quite sufficient for ordinary life physics.

2.10 Different Water Bodies and their Modeling

Rivers and Streams

The main focuss of water quality modeling has remained on river ever since its inception. The main reason for this is the importance of rivers for water supply and wastewater discharge. Apart from that it has got agreat aesthatic value.

Further the role of river and streams in Agriculture and navigation can hardly be underestimated. Moreover in a country like India rivers (Fig. 2.1) are considered sacred and provide site for religious expression of people.

Rivers can be classified in many ways. In the present context we may restrict ourselves to hydraulic i.e. flow, velocity, dispersion and geometric i.e. depth width and slope parameters which are essential for modeling them. It may be noted that flow pattern will vary according to monsoon season, glacier melting, topography, latitude etc. thus temporal variation inflow is also important. Further human interference like impounding, urbanization and consumptive use etc. affects the temporal variation greatly. However from the point of view of water quality modeling alone, warm low flow periods are most critical because of lesser purification capacity during this period. We will discuss this in detail as to how these parameters affects the modeling problem in Chapter-6. Here suffice is to saythat discharge, velocity and surface roughness are important factors for modeling them.

Fig. 2.1 Typical River Ecosystems

Estuaries

An estuary is a semi enclosed coastal body of water with one or more rivers or streams flowing into it, and with a free connection to the open sea (Fig. 2.2). These are regions of interaction between rivers and near-shore ocean waters, where tidal action and river flow mix fresh and salt water.

Fig. 2.2 Typical Coastal Ecosystems

Such regions include bays, mouths of rivers, salt marshes, and lagoons. Estuaries are often associated with high rates of biological productivity.These brackish water ecosystems shelter and feed marine life, birds, and wildlife.

The model is process based. The dominant physical, biological and human-related processes in an estuary are assumed. These processes are briefly summarised below.

The ecological health of estuaries is determined by the interaction between organisms and variations in salinity, currents, waves, Suspended Particulate Matter (SPM), bed sediments, temperature, air exposure, hypoxia, wetland contaminants and biodiversity. Like the health of a living organism, the health of an estuary or a coastal water body, cannot be measured by one single variable, indeed a number of variables are important. Well-flushed estuaries are intrinsically more robust than poorly flushed systems. As a result, environmental degradation is most often apparent during periods of reduced freshwater inflows, e.g. during drought or when human activities reduce the freshwater flow.

Once riverine derived suspended particulate matter enters the estuary, it can be trapped within an Estuarine Turbidity Maximum (ETM) zone. The ETM is commonly located in the very low salinity reaches of an estuary. The maximum, depth-averaged, Suspended Solid Concentration (SSC) at high water within an estuary can be predicted semi-empirically as a function of the tidal length and the tidal range.

Sediment particles and aggregates within the ETM can give rise to marked changes in water quality. Fine particles can adsorb metal ions and organic macro molecules from solution to such an extent that some metals can be completely removed from solution within a strong ETM. Once nutrients enter an estuary, non-conservative behaviour can be pronounced. Key processes responsible for this non conservative behaviour include burial in sediment reservoirs and desorption processes particularly if the sediment is nutrient-rich. Nutrients are generally mainly in particulate form (*i.e.* absorbed to the mud particles in suspension) in freshwater and can be released in solution in saline water.

The salt marshes of Western Europe generally produce more than 1 kg m^{-2} yr^{-1} of above ground dry matter. Salt marshes export some of this organic matter. Salt marshes and their tidal creeks are also an important nursery ground and a refuge for larvae and post larvae of bivalve, fish and zooplankton.

The estuary is modeled as a converter of living phytoplankton to detritus particles; it is also a conveyor of detritus matter to the sea. Fishes help transfer energy and matter from estuarine plants to upper trophic levels. The great bulk of the organic matter produced (sometimes 90%) is processed through the detritus system. Zooplankton, planktivorous fish, interstitial micro and meiofauna, surface deposit-feeding molluscs, fishes and polychaeta, and filter-feeding invertebrates consume a much greater proportion of the primary production of the phytoplankton and benthic microalgae. Annual plant growth and decay provides continuing large quantities of organic detritus. In addition, there is often a considerable input of detritus from river inflow. Detrital particles and their associated microorganisms provide the basic food source for primary consumers such as zooplankton, most benthic invertebrates and some fishes. The first trophic level in the estuarine ecosystem is therefore best described as a mixed trophic level of detritus consumers, which in varying degrees are herbivores, omnivores or primary carnivores.

In the model, freshwater phytoplankton and bacterioplankton in the river are subject to salt stress when freshwater mixes with saltwater; and the freshwater microbial populations die in this. In the model, the salinity also limits the seaward distribution of saline water plankton, invertebrates (e.g. bivalves) and fishes. Detail on water quality modeling and governing equation of them have been discussed in chapter-7 for coastal ecosystems.

Lakes and Reservoirs

In contrast to flowing water systems like rivers and estuaries they received comparatively less emphasis in initial years of modeling mainly because of relative less importance. However in recent years their importance has grown considerably. From the standpoint view of water quality modeling, we may classify lakes on the basis of origin, i.e. whether they are natural or man made; shape i.e. circular, elongated etc. and retention time (Fig. 2.3). Again it will need data on bathymetry i.e depth contours, pollutant load magnitude and location etc. for modeling are to be discussed in chapter-8.

Fig. 2.3 Typical Lakes Ecosystems

Oceans and Seas

Oceans are saline waters that cover almost three quarters (71%) of the surface of the Earth (Fig. 2.4).

Fig. 2.4. Oceans and Seas

The area of the oceans is 361 million sq. km., and nearly half of the world's marine waters are over 3,000 meters (9,800 ft) deep. The average salinity of the earth's oceans is around 35 parts per thousand (ppt), and nearly all seawater falls in the range of 31 to 38 ppt.Though somewhat arbitrarily divided into several "separate" oceans, these oceans are in fact one global, interconnected body of salt water, often called the World Ocean. The major divisions are defined in part by the continents and a variety of archipelagos, and are labeled the Pacific Ocean, the Atlantic Ocean, the Indian Ocean, the Southern Ocean, and the Arctic Ocean. Smaller regions of the oceans are called seas, gulfs, straits and other names.

There are also some smaller bodies of salt water that are not interconnected with the World Ocean. These are not considered to be oceans or parts of oceans, though some of them have been given the name, sea.

For modeling sea and ocean, bathymetric data, i.e. depth contours, current data, wind data, tide data are essential thus making modeling these region a rather complex phenomena. These are significantly used to know oil spill, tsunamis etc. which are explained in Chapter 9.

2.11 Factors Governing Water Quality Model

The hydrodynamic module and water quality module simulate water level variations, flows and pollutants transport respectively in response to a variety of forcing functions in lakes, estuaries and coastal regions. The effects and facilities include:

1. Bottom Shear Stress
2. Wind Shear Stress
3. Barometric Pressure Gradients
4. Coriolis Force
5. Momentum Dispersion
6. Sources and Sinks
7. Evaporation
8. Flooding and Drying
9. Wave Radiation Stresses

It can be applied to a wide range of hydraulic and related phenomena. This includes:

1. Modelling of Tidal Hydraulics
2. Wind and Wave Generated Currents
3. Storm Surges
4. Secondary Circulations, Eddies and Vortices
5. Harbour Seiching
6. Dam-Break
7. Tsunamis

The advection/ dispersion module provides the advection and dispersion of dissolved substances subject to advection and dispersion processes in lakes, estuaries and coastal regions. The effects and facilities include:

1. Linear decay
2. Heat dissipation to the atmosphere

This module can be applied to a wide range of hydraulic and related phenomena. This includes the spreading of dissolved substances such as Salt, Heat, Coliform bacteria, and Zenobiotic compounds. Typical applications include cooling water recirculation and water quality studies.

Mud Transport Module describes erosion, transport and deposition of mud or sand/mud mixtures under the action of currents and waves. It is applicable for:

1. Mud fractions alone, and
2. Sand/mud mixtures

The following processes can be included in the simulation.

1. Forcing by waves
2. Sliding
3. Salt-flocculation
4. Detailed description of the settling process
5. Layered description of the bed, and
6. Morphological update of the bed

In the Mud Transport Module, the settling velocity varies, according to the salinity, if included, and the concentration taking into account flocculation in the water column. Furthermore, hindered settling and consolidation in the fluid mud and under consolidated bed are included in the model. Bed erosion can be either non-uniform, i.e. the erosion of soft and partly consolidated bed, or uniform, i.e. the erosion of a dense and consolidated bed. The bed is described as layered and characterised by the density and shear strength.

The Mud Transport Module can be applied to the study of engineering problems such as:

1. Sediment transport studies for fine cohesive materials or sand/mud mixtures in estuaries and coastal areas in which environmental aspects are involved and degradation of water quality may occur.
2. Siltation in Harbours, navigational fairways, canals, rivers and reservoirs.

This chapter briefly elaborated about modeling basics in various water ecosystems with qualitative and quantitative parameters and their governing equations.

References

Canter, L. W., 1977. Environmental Impact Assessment, McGraw-Hill Book Company, Inc., New York, New York, Ch. 5, pp. 86-119.

Canter, L. W., 1985. Environmental Impact of Water Resources Projects, Lewis Publishers, Inc., Chelsea, Michigan, Ch. 2, pp. 77-110.

Interim Mekong Committee. Environmental Impact Assessment Guidelines for Application to Tropical River Basin Development, 1982, Mekong Secretariat, c/o ESCAP, United Nations Building, Bangkok, Thailand.

O'Connor, D.J., 1962. Organic pollution of New York Harbor - Theoretical Considerations. *J. WPCF*, 34(9), pp 905-919.

Streeter H W and Phelps EB (1925) A study of the pollution and natural purification of the Ohio river, III. Factors concerned in the phenomena of oxidation and reaeration. Bulletin 146. Public Health Service, Washington, DC, USA.

Thomann R.V.: 1963 "Mathematical Models for Dissolved Oxygen", *Journal of Sanitary Engineering*, ASCE, 89, 5.

❑❑❑

Chapter 3

Modeling: Processes and Development

3.1 Introduction

By the end of the twentieth century one of our most important scientific accomplishments has been the explicit realization that the human population is changing the earth system. Some of these changes are occurring at a remarkable rate. We know, with certainty, that human activities are changing land use, habitats, the chemistry of the Earth's atmosphere and water, rates and balance of biogeochemical processes, and diversity of life on the planet (Vitousek *et al.* 1997). One prominent mode of human disturbance has resulted from activities that mobilize the nutrient elements nitrogen and phosphorus through land clearing, production and applications of fertilizer, discharge of human waste, animal production, and combustion of fossil fuels (Nixon 1995).

As a result of these activities, surface waters and ground waters throughout the world now have elevated concentrations of N and P compared to concentrations even in the middle of the 20th century. Our mobilization of N and P has accelerated the fluxes of these elements to coastal waters, and fertilization of coastal ecosystems is now a serious environmental problem because it stimulates plant growth and disrupts the balance between the production and metabolism of organic matter in the coastal zone. Coastal eutrophication is a recently recognized phenomenon (Nixon 1995), and scientific investigation of this human disturbance has progressed for only a few decades, so our conceptual model of the problem is evolving rapidly. Here, I describe 2 phases in the evolution of our conceptual model of how nutrient enrichment disrupts coastal ecosystems, and then finish with a view to the future, suggesting critical questions that must be answered as we work to advance and broaden our understanding of this complex environmental problem. This examination of conceptual models is designed as a complement to the excellent general reviews of coastal eutrophication (Vollenweider 1992, Heip 1995, Nixon 1995, Jørgensen & Richardson 1996, National Research Council 2000).

The increasing N concentration are representative of changes in the nutrient (N and P) chemistry of rivers throughout the developed world, with many showing progressive increases that accelerated during the period from 1960 to 1990. The underlying mechanisms of these trends are well established (Puckett 1995, Jaworski et al. 1997, National Research Council 2000) and directly related to human activities that mobilize and accelerate the movement of N and P from the lithosphere to the hydrosphere. Our concern about this human disturbance is first motivated by the degradation of freshwater ecosystems by nutrient enrichment, and the study of lake, river eutrophication focus of freshwater ecological and geochemical research by the 1960s.

Nixon (1995) reminds us that this topic is not a focus of estuarine or coastal marine research then, so there is a lag of perhaps a decade or two before the problem of coastal eutrophication became established as a societal and scientific concern. It is natural, then, that the earliest conceptual model of coastal eutrophication is strongly influenced by the conceptual models developed first by limnologists. As we will see, there are fundamental differences in the system level responses to nutrient enrichment in lakes (Fig. 3.1) compared to estuarine-coastal ecosystems. One result of human enrichment of N and P in rivers is that the delivery of these elements to the coastal zone has increased proportionately. For example, Jaworski (1997) estimate that human activities have increased N fluxes to coastal rivers of the northeastern United States 5 to 14 times above natural rates. Phosphorus loading to estuarine systems has increased 2 to 6 fold since 1984 (Connolly, J.P. and R.P. Winfield, 1984.). Among those estuaries and coastal marine ecosystems in which there have been sustained observational programs, most have revealed significant decadal scale trends of increasing Nitrogen and Phosphorus availability.

Fig. 3.1 Typical Ecological System in the Nature

3.2 Modeling Processes

The water quality modeling processes is started with mass balance concepts. The following figure (Fig. 3.2) shows the mass balance concept models for a typical model. It includes system boundary, external loading, transport in and out substance transport out etc. Initially, the modeler can be derived the mass balance equations and variables within the system boundary.

The modeling process of water quality is shown in figure 3.3 with different variables such as upstream loading , aquatic animals, uptake from soil and bottom, air water exchange etc.. Some variables are received to system and some variables are releasing from the system. It is observed from the figure 3.4, the reaction with in the system also considered in the concept model.

Figure 3.5 and 3.6 shows the bioaccumulation of toxic concentration and system components in various trophic levels in the water ecosystems such as primary producers, consumers and dead materials in the ecosystems.

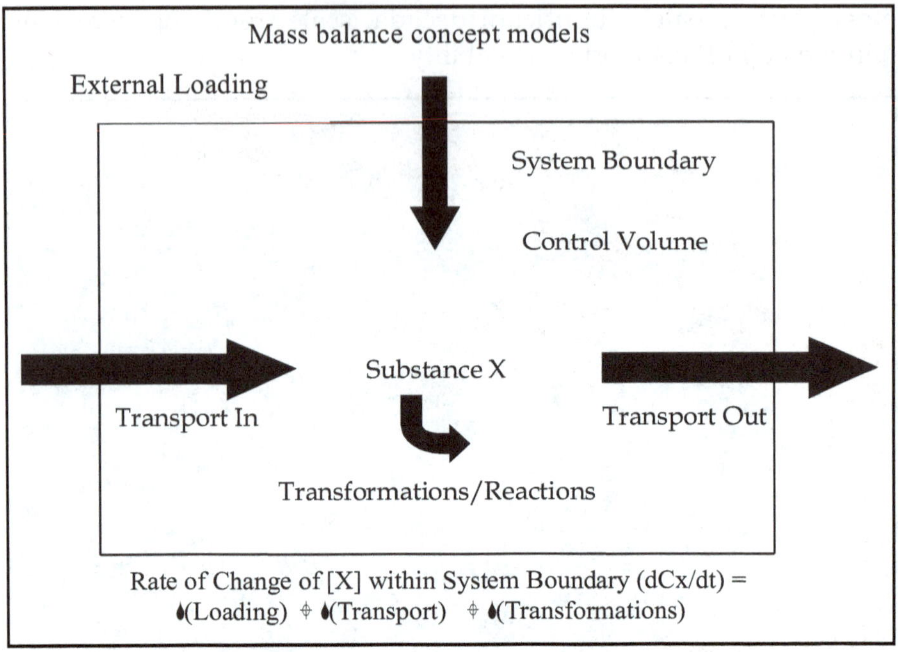

Fig. 3.2. Typical Mass Balance Concept Models

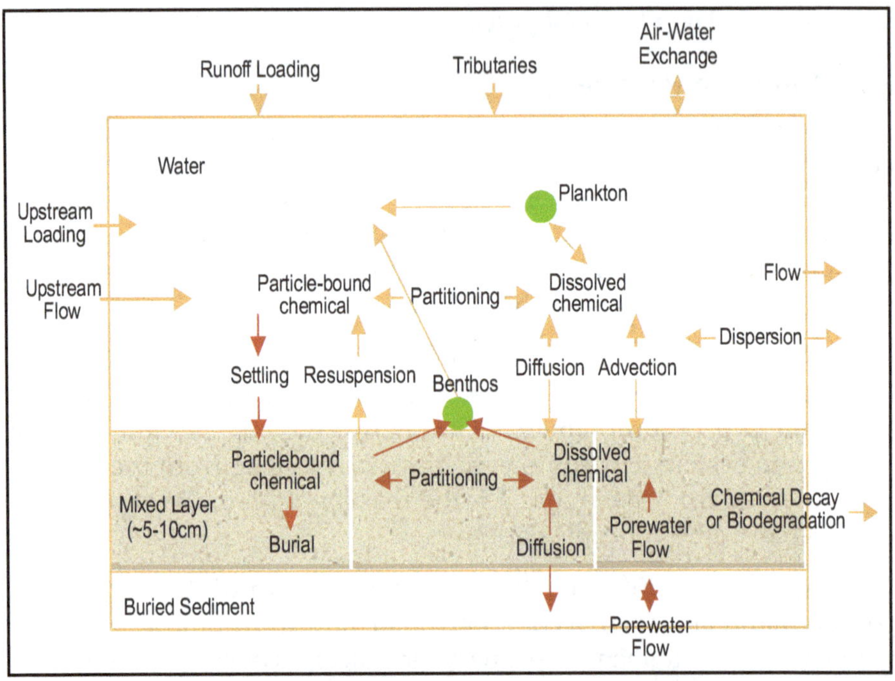

Fig. 3.3 Example Exposure Model Framework

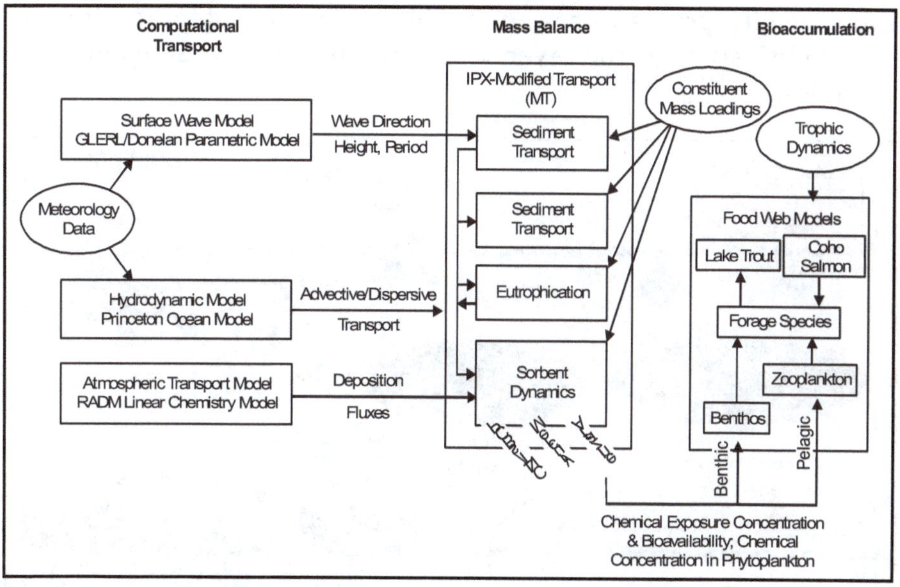

Fig. 3.4 Lake Mass Balance Study Model

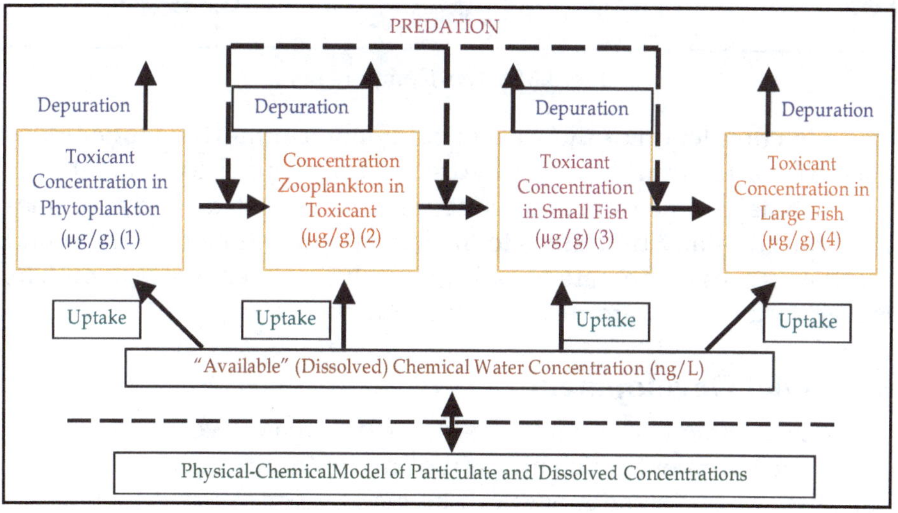

Fig. 3.5 Bioaccumulation Model Framework

3.3 Model Architecture

Following the initialization of the model and the reading of fixed data (e.g. hydrography), data for the first day of the simulation are read. This allows for the calculation of the sub-daily timestep. The hydrodynamic processes, the particle settling processes, and the water quality processes are advanced on this time step, resulting in the modification of the layer structure and of the values of layer variables.

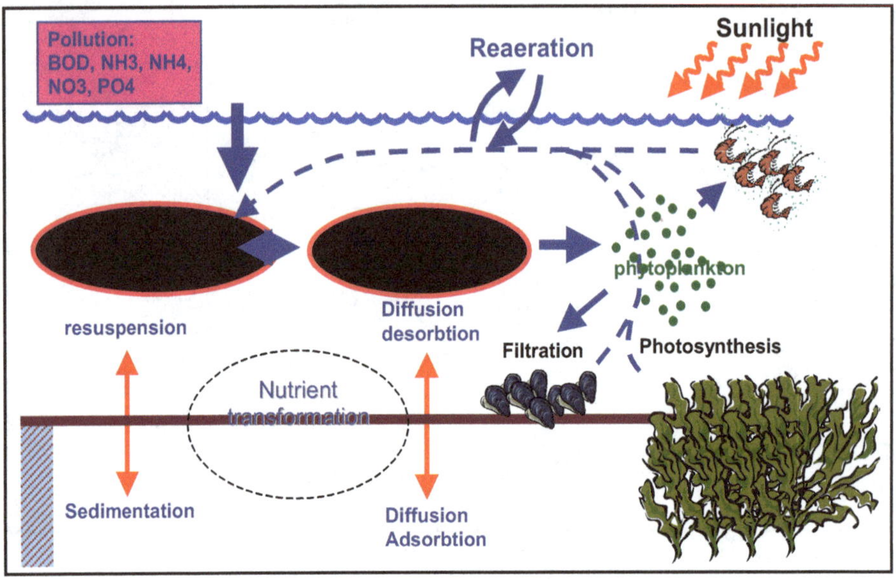

Fig. 3.6 System Components

On completion of the water quality algorithms, the model loops back to the beginning of the hydrodynamic model to calculate the next sub-daily time step, and the sequence is repeated. This continues until the sub-daily time steps total 24 h, at which time the inflow and outflow processes operate for a single 24 h time step. Once complete, the next day's data are read and the entire process repeated.

3.4 Model Development

Development of model consists of two major works the first is theory development which means writing continuity equation for mass balances. At this stage model can be made a simpler one such as 1D and 2D and ignoring some could not important parameters or may be a complex one such as 3D with overall consideration to all parameters for example wind, wave radiation, stress are also considered along with other major parameters like tide bathymetry in our coastal environmental models.

Some of the key factors that helps with the whose model is more suitable are

1) Cost consideration
2) Objectives of studies
3) Accuracy needed
4) Time varying

The second stage in programme development. This stage course of developing the actual numerical algorithm in one or more computer language. A range of facts and performed to know whether the model is functioning according or not. After obtaining the numerical relations the results are plotted by using same or different programme.

3.5 Governing Equations

A stationary sediment bed may be validly assumed in reservoirs and lakes where wind has little effect on bottom conditions. As toxic substances appear to decay very slowly (first-order rate constant of 0.001–0.1 d^{-1}) in adsorbed particulate form or sediment environment (Schnoor, 1996; Thomann and Mueller, 1987; Di Toro *et al.*, 1981), sediment decay rates of toxic chemicals are assumed negligible compared with those in the dissolved phase. If sediment decay rates are significant, laboratory or field data are required to accurately determine the rates. Linear sorption–desorption kinetics were adopted because chemical concentrations in reservoir water are mostly environmentally relevant. An instantaneous local equilibrium is assumed for sorption processes because the time scale for sorption reactions is much smaller than that of other kinetic and macroscopic transport (advection and diffusion) processes in reservoirs. With the aforementioned assumptions, total chemical concentrations in water column ($C_{t,w}$) and bed sediments ($C_{t,b}$) are formulated by the conservation of mass:

$$\frac{\partial(BC_{t,w})}{B\partial t} + \frac{\partial(UBC_{t,w})}{B\partial x} + \frac{\partial(WBC_{t,w})}{B\partial z} - \frac{\partial}{B\partial x}\left(BD_x\frac{\partial C_{t,w}}{\partial x}\right) - \frac{\partial}{B\partial z}\left(BD_z\frac{\partial C_{t,w}}{\partial z}\right)$$

$$= \frac{K_f}{y}(f_{d,b}C_{t,b/\phi} - f_{d,w}C_{t,w}) - (K_p + K_H + K_O + K_B)f_{d,w}C_{t,w} \quad \text{..... (3.1)}$$

$$+ \frac{k_l}{z}[(C_a/H) - f_{d,w}C_{t,w}] - \frac{U_s}{z}\, fp,w\,Ct,w + \Phi NPS$$

$$\frac{\partial(BC_{t,b})}{B\partial t} = -\frac{K_f}{y}(f_{d,b}C_{t,b}/\phi - f_{d,w}C_{t,w}) - (K_H + K_O + K_B)f_{d,b}C_{t,b} + \frac{U_s}{\Delta z}f_{p,w}C_{t,w} \quad \text{..... (3.2)}$$

where subscripts t, d, and p denote the total, dissolved, and particulate phases of a chemical, respectively; subscripts a, w, and b denote air, water, and bed, respectively; f_d and f_p are the fractions of dissolved and particulate chemicals to the total chemical; t is time; x is longitudinal Cartesian coordinate (positive to the right); B is waterbody width; U and W are longitudinal and vertical flow velocities, respectively; D_x and D_z are longitudinal and vertical constituent dispersion coefficients, respectively; K_f is diffusive exchange rate between water column and pore water of the bed; ϕ is the porosity of the bed sediments; K_P, K_H, K_O, and K_B are rate constants for photolysis, hydrolysis, oxidation, and biotransformation, respectively; k_1 is air–waterexchange rate due to volatilization; H is Henry's constant; C_a is vapor phase concentration; $_{NPS}$ is the lateral mass flow rate of constituent per unit volume; v_s is the net settling velocity of sorbed chemical; z is the depth of water from water surface; and y is the distance from reservoir bottom.

The chemical kinetic reaction rates and model input parameters can be determined from field and laboratory experiments, estimation using chemical properties, and from the literature (Lyman *et al.*, 1982; Schnoor *et al.*, 1987).

Transport of sediments in the water column and reservoir bed is formulated using the two-dimensional laterally integrated advection–diffusion equations, assuming a stationary bed condition:

$$\frac{\partial(BC_s)}{B\partial t}+\frac{\partial(UBC_s)}{B\partial x}+\frac{\partial(WBC_s)}{B\partial z}-\frac{\partial}{B\partial x}\left(BD_z\frac{\partial C_s}{\partial x}\right)-\frac{\partial}{B\partial z}\left(BD_z\frac{\partial C_s}{\partial z}\right)= -\frac{Uss}{\Delta z}C_{ss}-K_{dt}\gamma om\,C_{dt}-\frac{U}{\Delta z}C_{dt}+q_1 \quad \text{..... (3.3)}$$

$$\frac{\partial(BC_b)}{B\partial t}=\frac{U_{ss}}{\Delta z}Css+\frac{U_{dt}}{\Delta z}C_{dt}-\gamma om\,K_sC_b+q_b \quad \text{..... (3.4)}$$

where C_s is sediments concentration including inorganic and organic (detritus) sediments in water column; q_1 is lateral mass flow rate of sediments per unit volume for water column; v_{ss} is net settling velocity of suspended solids; C_{ss} is suspended solids concentration; Δz is grid cell thickness; K_{dt} is detritus decay rate; v_{dt} is net detritus settling rate; C_{dt} is detritus concentration in water column; C_b is the total bed sediment concentration; γ_{om} is rate multiplier for temperature adjustment; K_s is organic bed sediment decay rate; and q_b is lateral mass flow rate of sediments per unit volume for bed sediments.

3.6 Physical, Chemical and Biological Processes

A partition coefficient (*K*) is used to determine the fractions of dissolved and particulate chemical forms to the total chemical based on the linear sorption-desorption kinetics. For the equilibrium phase partitioning, two phases and particle interaction models (Di Toro, 1985; Bierman, 1994) are adopted. In the initial stages of a chemical spill, dissolved chemical in the water column may transfer to interstitial water in the bottom sediment by diffusion processes, whereas it may be diffused from the bottom sediment to the water column during the recovery phase, depending on the concentration gradient. This mass transfer is estimated as a function of molecular weight (*M*) of a chemical and porosity (ϕ) of bed sediments (Di Toro *et al.*, 1981) as follows:

$$K_f = 19\phi(M)^{\frac{-2}{3}} \qquad \text{..... (3.5)}$$

The amount of chemical loss due to photolysis is a function of the quantity and wavelength distribution of incident light, the light absorption characteristics of the chemical, and the efficiency at which absorbed light yields a chemical reaction. A first-order reaction constant was used to estimate the rate of photolysis (Thomann and Mueller, 1987):

$$K_P = K_{do} \frac{I_0}{I_0^1} \frac{D}{D_0} \left\{ \frac{1 - \exp[-K_e(\lambda_{max})z]}{K_e(\lambda_{max})z} \right\} \qquad \text{..... (3.6)}$$

where K_{do} is the direct near-surface photolysis rate; I_o is the daily amount of incoming solar radiation at the water surface; I_o' is the light intensity at which K_{do} was measured; *D* is the radiance distribution function; D_o is the radiance distribution function near the surface; and $K_e(\lambda_{max})$ is the net light extinction coefficient at λ_{max}, the wavelength of the maximum light absorption. Extinction coefficients for water, inorganic, and organic sediments were used to calculate a net light extinction coefficient, K_e (Cole and Buchak, 1994).

Hydrolysis is a major dissipation process in which a chemical compound reacts with water molecules and results in a cleavage of chemical bond. Generally, hydrolysis is a second-order reaction because of dependence on the molar concentrations of $[H^+]$, $[OH^-]$, or water mediators. A pseudo-first order rate constant, $K_H = K_n + K_a[H^+] + K_b[OH^-]$, is used, *where* K_n is the neutral hydrolysis rate, K_a is the acid catalyzed hydrolysis rate, and K_b is the base catalyzed hydrolysis rate. The Arrhenius function was used to adjust the rate with temperature (Thomann and Mueller, 1987).

Toxic chemicals can be oxidized by a reaction with either dissolved oxygen or free radicals in natural waters (Schnoor, 1996). A pseudo-first order reaction rate constant is used to compute degradation of a chemical by oxidation assuming that the rate of free-radical formation is relatively constant. The rate is adjusted for water temperature with the Arrhenius equation.

Biotransformation is the microbially mediated decay process. The model is designed to treat the biotransformation process either by second-order or by pseudo–first order kinetic reactions based on the Monod equation:

$$\frac{dC_d}{dt} = -\frac{1}{yB}\frac{dC_B}{dt} = \frac{\mu_{max}}{y_B}\left(\frac{C_d}{K_x + C_d}\right)C_B \qquad \text{..... (3.7)}$$

where C_B is the bacterial concentration; y_B is the bacterial yield coefficient; μ_{max} is the maximum specific growth rate; and K_x is the half-saturation constant. The rate is adjusted for water temperature.

A modified two-film theory (Mackay, 1982; O'Connor, 1983) is used to compute the gaseous transfer of chemical from air to water and water to air. The transfer rate is proportional to the concentration gradient between the chemicals in the water column and in the overlying atmosphere. The overall transfer coefficient, k_l, is given as follows:

$$\frac{1}{k_1} = \frac{1}{K_1} + \frac{1}{K_g H} \qquad \text{..... (3.8)}$$

where K_l is the liquid film coefficient and K_g is the gas film coefficient. The k_l value is computed as a function of the chemical characteristics (H, K_l, and K_g), water velocity, and wind speed.

Since the transfer coefficient for the open bodies of water such as reservoir and lake are largely affected by wind, O'Connor (1983) and Mackay (1982) equations are incorporated into the model for the estimation of the liquid and gas film transfer coefficients.

3.7 Numerical Method and Computer Programming

The governing equations are solved using the finite difference method as in the laterally integrated hydrodynamics and mass transport model, CE-QUAL-W2 (Cole and Buchak, 1994). The dependent variables are toxicant and sediment concentrations in the water column and bed sediments. The independent variables are longitudinal distance, flow depth, and time. The flow chart in briefly describes the overall algorithms of the model. The water surface elevation and horizontal momentum are computed simultaneously in the model based on an implicit finite difference scheme, which allows the use of a reasonable time scale for field application over entire stratification cycles. The equations for toxicant and sediment transport are solved using the explicit QUICKEST (Quadratic Upstream Interpolation for Convective Kinematics with Estimated Streaming Terms) finite difference scheme (Leonard, 1979). Vertical turbulent transfer of toxic contaminants is determined from the vertical shear of horizontal velocity and a density gradient dependent Richardson number function (Chapra, 1980).

The computer program for the toxics submodel is created and linked to the CE-QUAL-W2 code by using the Microsoft (Redmond, WA) Fortran Powerstation program, which can be found in Chung (1998). The physical, chemical, and biological properties of a toxic chemical and its kinetic reaction rates need to be provided through an input file. The model can compute kinetic decay processes either independently by providing individual kinetic reaction rates or collectively by providing a lumped value of transformation rate or half-life. The values of half-life and its variation with time can be specified in the model.

3.8 Boundary and Initial Conditions

The model generates vertical and longitudinal distributions of a toxic chemical in response to time varying boundary conditions. Flows and contaminant loadings are specified at the upstream boundary and tributaries. The discharges of contaminant through outflow are used as downstream boundary conditions. At the surface of the waterbody, precipitation, evaporation, and meteorological changes over time are provided. The lateral boundary conditions include overland runoff, ground water gain or loss, water withdrawal, and concentrations of contaminants contained in the lateral flows. The time varying flows and mass loadings can be defined as either a step function or linearly interpolated value between two data points.

The initial toxic substance and sediment concentrations in the water column can be specified as either a uniform value, a single vertical profile for all segments, or a vertical profile for each segment. In bed sediment, they can be specified as either a uniform value or longitudinal profile for each segment. If nonuniform initial conditions are preferred, grid-wide vertical or longitudinal profiles should be provided in an input file.

A two-dimensional reservoir toxics submodel is developed and incorporated into a laterally integrated hydrodynamics and transport model. The model is capable of simulating the fate and transport processes of toxic contaminants, including sorption and desorption, photolysis, hydrolysis, oxidation, biotransformation, volatilization, diffusive exchanges between the bottom sediment and water column, and sediment transport and deposition in a reservoir. The development of the toxics model enhanced the versatility of the original reservoir model and provided an effective tool for describing and predicting the behavior of a toxic contaminant in a stratified reservoir during a spill or runoff.

3.9 Model Refinement

Model refinement of the hydrodynamic module and water quality module can be done with small adjustments in the following variables.

1. Bottom Shear Stress
2. Wind Shear Stress
3. Barometric Pressure Gradients
4. Coriolis Force
5. Momentum Dispersion
6. Sources and Sinks
7. Evaporation
8. Flooding And Drying
9. Wave Radiation Stresses

The advection/ dispersion module provides the advection and dispersion of dissolved substances subject to advection and dispersion processes in lakes, estuaries and coastal regions. The effects and facilities include:

1. Linear decay
2. Heat dissipation to the atmosphere

This module can be applied to a wide range of hydraulic and related phenomena. This includes the spreading of dissolved substances such as Salt, Heat, Coliform bacteria, and Zenobiotic compounds. Typical applications include cooling water recirculation and water quality studies.

3.10 Model Calibration

Model calibration process is mainly based on the field measurements that helps in choosing the empirical coefficients in water quality models and also in the verification of the consistency of the model's initial and boundary condition with that of the in-stream measurements. Calibration of the hydrodynamic part of the model is first carried out by comparing simulated hydrodynamic variables with the measured ones.

Next, the calibration of the process compartment of the model is carried out sequentially by using transformation kinetic parameters. The order of calibration is; temperature, BOD, DO, nitrification and phosphorous related parameters respectively. While calibrating water quality process part of the model, estimated parameters are fixed, few parameters are extracted from standard modeling literature, remaining parameters are obtained by tuning them till the observed and predicted results closely matched.

The calibrated model is used to predict the water quality with an independent set of data as a part of validation exercise. Results of model predictions are fairly good, and performance of the model is further confirmed through statistical evaluation of the results. Statistical analysis of the model results indicate that, simulated Carbonaceous Biochemical Oxygen Demand results are relatively in best agreement with the measured values with coefficient of regression 0.9, coefficient of variation 7.2%, RMS 19.2% and relative error 8%.

One dimensional steady state water quality model is calibrated and validated. Model coefficients are estimated through field and laboratory studies. Performance of the calibrated model is evaluated through statistical techniques. Model output is found to be very sensitive to headwater quality. Sensitivity of the Carbonaceous Biochemical Oxygen Demand decay rate is found to be increasing towards downstream of the outfall. Model coefficients related to settling rates of CBOD, Nitrogen and Phosphorous fraction are found to be moderately sensitive. Among all the model coefficients of the Qual-2e, dispersion constant is found to be least sensitive and seems to have been included only to take care of numerical dispersion. Based on statistical evaluation of the calibrated model it is evident that, the model has been calibrated reasonably well. Among the state variables presented here, prediction of CBOD results found to be better with relative error of 8%, considering the fact that 20 % deviation is normally acceptable in standard river water quality modeling practices. The water quality modeling processes and development described in this chapter with different architectures, calibrations, validation and their formulae.

References

Bierman, V.J., Jr. 1994. Partitioning of organic chemicals in sediments:Estimation of interstitial concentrations using organism body bur- comparison with the interflow regime, the overflow redens.p. 153–175. In J.V. DePinto, W. Lick, and J.F. Paul (ed.)

Chung, S.W. 1998. Modeling the fate and transport of agricultural pollutants and their environmental impact on surface and subsurface water quality. Ph.D. diss. Dep. of Civil and Construction Eng., Iowa State Univ., Ames.

Chapra, S.C., 1980. Total Phosphorus model for the Great Lakes. *J. Eng. div. Am. Soc. Civil Engr.* 103: 147-161.

Cole T.M., Buchak E. M. 1994 (CE-QUAL-W2) A two dimensional laterally averaged, hydrodynamic and water quality model, Version 2.0. User Manual. US ACE

Connolly, J.P. and R.P. Winfield, 1984. A user's guide for WASTOX, a framework for modeling the fate of toxic chemicals in aquatic environments. Part 1. Exposure concentration. USEPA Gulf Breeze, *FL. EPA-600*/3-84-077.

Di Toro, D.M., D.J. O'Connor, R.V. Thomann, and J.P. John. 1981.Analysis of fate of chemicals in receiving waters. Phase 1. Chem. Manufact. Assoc., Washington, DC.

Di Toro, D.M. 1985. A particle interaction model of reversible organic chemical sorption. *Chemosphere* 14:1503–1538.

Heip, C.H.R., Hemminga, M.A. and Bie, M.J.M. (eds) 1995 Monsoon and coastal ecosystems in Kenya. Netherlands Indian Ocean Programme. *Netherlands Museum of Natural History Cruise Report* No. 5.

Jaworski, N. 1997. Total nitrogen (TN) and phosphorus (TP) loadings and TN:TP molar ratios for the estuaries of the Middle Atlantic and Northeast U.S.A. during the last century. *Estuarine Research Foundation Conference.*

Jorgensen, B. B., And K. Richardson [Eds.]. 1996. Eutrophication in coastal marine ecosystems. American Geophysical Union, Coastal and Estuarine Studies 52, Washington, D.C. ISBN 87590-266-9. 273 p.

Leonard, B. P. 1979. A Stable and Accurate Convective Modeling Procedure Based on Upstream Interpolation. *Computer Methods in Applied Mechanics and Engineering*, 19, 59–98.

Lyman, W.J., W.F. Reehl, and D.H. Rosenblatt. 1982. Handbook of Chemical Property Estimation Methods. McGraw-Hill, New York.

Mackay, D. 1982. Volatilization of organic pollutants from water. *EPA/600/3-82/019.* USEPA, Athens, GA.

NRC (National Research Council). 2000. Toxicological effects of methylmercury. Committee on the Toxicological Effects of Methylmercury. National Academy Press. Washington, DC.

Nixon, S. W. (1995). Coastal marine eutrophication - A definition, social causes, and future concerns. *Ophelia*, 41, 199–219.

O'Connor, D.J., 1983. Wind effects on gas-liquid transfercoefficients. *J. Environ. Eng.* 109, 731–752.

Puckett, L.J. (1995), Identifying the Major Sources of Nutrient Water Pollution, Environmental Science & Technology, 29, 408A-414A.

Schnoor, J.L. 1996. Environmental modeling: Fate and transport of pollutants in Water, Air, and Soil. John Wiley & Sons, New York.

Schnoor, J.L., C. Sato, D. McKechnie, and D. Sahoo. 1987. Processes,coefficients, and models for simulating toxic organics and heavy metals in surface waters. *EPA/600/3-87/015.* USEPA, Athens, GA.

Thomann, R.V. and J.A. Mueller. 1987. Principles of Surface Water Quality Modeling and Control. New York: Harper & Row.

Vitousek, P.M., Aber, J.D., Howarth, R.W., Likens, G.E., Matson, P.A., Schinlder, D.W., Schlesinger, W.H., and Tilman, D.G., 1997, Human alteration of the global nitrogen cycle—sources and consequences: *Ecological Applications*, 7, p. 737-750

Vollenweider, R. A., Marchetti, R. and Viviani, R., Eds. (1992). Marine Coastal Eutrophication. The Response of Marine Transitional Systems to Human Impact: Problems and Perspectives for Restoration. Science of the Total Environment Supplement. Amsterdam, Elsevier Science.

Chapter 4

Modeling: Applications

4.1 Introduction

The developed or commercially available models can be applied for any water bodies such as fresh water lakes, ponds, rivers, estuaries and marine ecosystems. The predicted results can be compared with proved with field data and existing approved models and applies to the other such ecosystems. In this chapter, we can discuss about modeling of water bodies, water quality model, spatial discretizafion, modeling pollutants and toxicants, concentration modeling, hydrodynamic mixing zone analysis.

4.2 Modeling of Water bodies

The need for predictive water quality modeling has arisen largely as a result of increased eutrophication of lakes throughout the world (Forsberg, 1987; Canfield and Hoyer, 1988). The most common modelling approach is exemplified by the development and application of steady state, input-output models. Generally, nutrient concentrations are calculated from net inputs and chlorophyll a concentration is predicted by correlation with the limiting nutrient, most often phosphorus (Dillon and Rigler, 1975; Canfield and Bachmann, 1981; OECD, 1982).

Factors that can also influence phytoplankton biomass, such as light, climate, biological interactions and internal loading of nutrients, are not considered. Further more, the assumption that a lake is a continuously mixed system is very restrictive and only applicable at discrete times of the year. As a result, the shortcomings of such approaches include an inability to make predictions in the face of varying physical and biological conditions, and a failure to offer insights into the determinants of changing water quality.

A second approach, often referred to as ecological water quality modeling, specifically addresses many of the biological and chemical factors that are absent in the simple input-output models. This approach has evolved in order to obtain a more fundamental understanding and representation of the major physical, chemical and biological processes that affect the biomass of phytoplankton and higher tropic levels (Di Toro *et al.*, 1971; Scavia, 1980; J~rgensen, 1983; Matsuoka *et al.*, 1986). Typically such models represent ecological processes by time varying, interdependent conservation equations, with rate coefficients that require calibration to suit site-specific conditions. The need for site-specific calibration is an undesirable feature of these models. However, present understanding does not afford a more fundamental, calibration-free description of the range of processes covered. In ecological water quality models, the physical processes of transport and mixing within the water body have generally been oversimplified, with the assumption of a continuously stirred or two compartment vertical system being common. Interactions between physical processes and the biological and chemical processes described by these models are represented.

The predictive abilities of these models are therefore compromised, despite increased ability to simulate the response to a range of altered biological and chemical conditions. A third approach has been the extension of hydrodynamic models to include water quality components, either by combination with simple input-out put models or more recently with ecological models. Orlob (1983) and Watanabe *et al.* (1983) review the evolution of such models. The most common approach for the hydrodynamics has been to use a one-dimensional (1D) model, with retention of variations in the vertical dimension. Many of the early, widely used models are not process based, and instead solve the 1D, thermal advection-diffusion equation (Dake and Harleman, 1966; Orlob and Selna, 1970). The calibration required for the hydrodynamic model, when coupled with the calibration of the biochemical component of the model, is very limiting and the results are of questionable value when predictions are required under conditions of changed inputs.

Three process based, hydrodynamic models are those of Stefan and Ford (1975), Bloss and Harleman (1980), and Imberger et al. (1978). The first two have been coupled with ecological models, giving the water quality models MINLAKE (Riley and Ste fan, 1987; Riley and Stefan, 1988) and CE-QUAL (USCE, 1986) respectively. All three hydrodynamic models present a description of the mixing and transport processes associated with inflow, outflow, diffusion and mixed layer dynamics. The models differ most in the extent to which the individual processes are described. The mixed layer dynamics, arguably the most important process in terms of algal production, provide the best example. Each model recognizes the production of Turbulent Kinetic Energy (TKE) as the basis of the description of mixed layer deepening. The Stefan and Ford model relates the total TKE production to the observed mixed layer deepening via an adjustable parameter. The Bloss and Harleman model divides the TKE production into two components, with one of these adjustable to allow for different mixing efficiencies under different levels of stratification and wind stress.

The Imberger *et al.* model, DYRESM, computes the TKE produced through four discrete processes. The amount of mixing is at a fixed rate for each process, having been determined independently from consideration of theoretical, laboratory and field results (e.g. Imberger and Patterson, 1981). The fact that only the Imberger et al. (1978) model is free of calibration implies that the level of process description, including the temporal and spatial scales in the model, is fundamentally correct. This result has been confirmed by the application of the model to a large number of lakes and reservoirs (Imberger and Patterson, 1990). DYRESM has therefore been used for the hydrodynamic component of the water quality model described below.

The purpose of the present chapter is to present a coupled hydrodynamic and ecological water quality model in which the hydrodynamic component is totally calibration. The hydrodynamics can be one of the dominant factors in lake ecology (Vincent et al., 1991), so a major advantage of such an approach is that it extends the range of application of the model to cover altered hydrodynamic conditions in which calibration data may exist. This approach also enables the interactions between ecological and hydrodynamic processes to be examined at a more fundamental level.

The DYRESM Water Quality model therefore represents a significant improvement over the few existing models of this type. It is still restricted, however, to calibration of the ecological algorithms. While there has recently been progress made towards mechanistic descriptions of ecological processes (e.g. Jorgensen, 1992; Kmet *et al.*, 1993; Patterson *et al.*, 1994) and in numerically quantifying the values of specific parameters (Jorgensen *et al.*, 1991) present understanding has not evolved to the extent that eutrophication models are complete calibration.

It should be followed an approach similar to that taken in the earlier coupled models, in that we rely on the ecological modeling literature to guide our selection of algorithms that describe the biological and chemical processes. Some of the individual choices described below are identical to those of MINLAKE or CE-QUAL, but the coupled water quality model is unique. In what follows we describe the hydrodynamic model, the particle settling model, and the ecological model that comprise the present water quality model. A description of the calibration procedure adopted for the ecological model, and a comparison between model predictions and field observations is presented in the following chapters.

4.3 Water Quality Model

The model, DYRESM Water Quality (DWQ), comprises a number of components. The hydrodynamic component includes separate algorithms for individual mixed layer processes, inflow (both riverine and groundwater), natural or man-made outflows, and hypolimnetic mixing. The particle settling component allows for the gravitational settling of a spectrum of particle sizes with variable degree of coagulation. The ecological component models phytoplankton production, nutrient cycling, and the dissolved oxygen budget. The following sections describe these components.

As some components have been well documented previously, their descriptions will be brief. However, the description is considered necessary to explain the functioning of DWQ. Descriptions are also given of the model architecture, time step, spatial discretization, input data requirements, and output.

4.3.1 Spatial discretization

The spatial discretization is identical to that used in the 1D hydrodynamic model DYRESM. The vertical profile of the lake is represented as a set of up to 100 Lagrangian layers which are free to move vertically, and to contract and expand in response to inflows, outflows and surface mass fluxes. The Lagrangian formulation avoids the need to calculate vertical velocities, greatly decreasing computational time and minimizing numerical diffusion. Each layer is homogeneous, and property differences between layers represent the vertical distribution. Layer thick nesses are adjusted within the model according to the resolution required to represent the vertical density gradient. The model is initialized with measured data for each layer. Properties of the amalgamated layer are volumetrically averaged, and the total number of model layers is decreased accordingly. Amalgamated layers may be split according to a specified maximum layer thickness criterion. Conversely, when a layer size falls below a specified minimum criterion, as may occur when there is withdrawal of water from the lake, then the layer is amalgamated with the smaller of the two bounding layers.

A unique feature of the spatial discretization is an extension to allow a quasi two dimensional representation of riverine inflows (Jokela and Patterson, 1985), as it may take several days for inflows to intrude across the full length of larger lakes. It is only when the inflow for a particular day has completely crossed the lake that water from that inflow is added to the 1D layer structure.

4.4 Modeling Pollutants and Toxicants

The use of chemicals such as pesticides, herbicides and insecticides in the field and industrial effluents have produced various environmental problems. Accidental spills or storm runoff of toxic substances into rivers and reservoirs can severely affect fisheries and local water supply systems in a short time period ranging from hours to days (Capel *et al.*, 1988; Chatterjee, 1991). The environmental and economic effects of a chemical spill into a reservoir can be fatal for a community, in particular if the community is heavily dependent on the reservoir for water supply. In Midwestern U.S. reservoirs, frequent occurrence of high levels of pesticides during late spring and early summer is one of the most serious water quality issues (Goolsby *et al.*, 1993). The levels of some pesticide concentrations occasionally exceeded their Maximum Contaminant Level (MCL) for drinking water with potential adverse effects on human health (Eldridge *et al.*, 1994). Therefore, a full understanding of the transport and fate of toxic substances in a reservoir during a spill or peak runoff loading period is essential to establishing an effective management strategy and protecting the scarce water resources.

Once a toxic chemical spill or runoff enters a reservoir, its fate and transport processes are governed by various factors including flow conditions, chemical and biological conditions of the water body, weather factors, and properties of the toxicant. The processes can be further complicated if the reservoir is stratified and a density current forms due to the temperature difference between river and ambient waters. The kinetic processes can be largely influenced by flow regimes. On entering the reservoir, river flow pushes water ahead to some distance until the buoyancy force, attributable to the density difference between the river and reservoir surface water, becomes sufficient to counteract the momentum of incoming water. An overflow occurs when the density of the river inflow is less than or identical to that of the receiving reservoir water, moving forward over the surface of the reservoir.

The negative buoyancy force drives the inflow downward to sink below the surface when the river water is heavier than the ambient reservoir water, forming a plunge flow. After plunging, the river flow is submerged down the old river channel and moves along the riverbed, becoming an underflow. Both plunge flow and underflow are induced by initial momentum and negative buoyancy due to density difference. At a distance where the density of the underflow is reduced by mixing and dilution to that of the ambient reservoir water, the underflow separates from the bottom and penetrates horizontally into the reservoir, forming an intruding interflow. Underflow and interflow are controlled by reservoir stratification and mixing with ambient water.

When the inflow becomes an overflow, some chemicals dissipate more readily through volatilization, photolysis, and oxidation because sufficient turbulence, sunlight, and oxygen are available near the water surface. Therefore, a water quality model should adequately link the kinetic processes of a toxic chemical with reservoir hydrodynamics for better prediction of the persistence and spatial distribution of the toxicant during a spill or runoff.

A laterally integrated two-dimensional hydrodynamics and transport model has been used for modeling water temperature and conventional water quality constituents in reservoirs (Martin, 1988; Cole and Buchak, 1994). Chung and Gu (1998) employed the model to simulate and analyze the transport of a toxic pesticide that spilled into a reservoir by treating the toxicant as a conservative substance. Due to the lack of a toxic contaminants modeling component, the model's application has been limited to the transport and mixing processes of a tracer under certain flow regimes such as a deep interflow, where the kinetic processes of the toxic substance may be insignificant. To understand the effectiveness of physico-chemical dissipation processes and behavior of the toxic contaminant under various flow regimes, the development of a toxics modeling component is necessary.

In this book, to help better understand the kinetic processes and persistence and predict the dissipation of toxic contaminants in a water ecosystems, a two-dimensional reservoir toxics model is developed and tested against field data. The aim of this modeling investigation is to provide information useful in the analysis of contaminated flows in a stratified reservoir. The model was constructed by developing a toxics sub model and incorporating it into an existing reservoir hydrodynamics and transport model. It simulates unsteady vertical and longitudinal distribution of a toxic chemical with respect to various transport and transformation processes.

The model is validated against field data collected from the Shasta Reservoir, California, during the pesticide VAPAM spill after railroad tank cars derailed on a canyon bridge over the Sacramento River in 1991. Observed and simulated chemical concentrations are employed to analyze the fate and transport processes of a product of the spilled chemical, methyl isothiocyanate.

Numerical experiments, using computational runs for selected scenarios, are conducted to examine the effect of two reservoir flow regimes, interflow and overflow, on the dissipation of the volatile toxic compound in the reservoir during the spill event. The analysis and results of this study are intended to assist in contamination control and remediation after a toxic spill and guide field monitoring and water quality management.

4.5 Concentration Modeling

The numerical software can be used to simulate dispersion of concentrated saline and hot water in river, coastal, seas, estuary, lake, streams ecosystems. It is based on the following non-linear vertically integrated two dimensional equation of conservation of mass which takes into account advection, dispersion. The following equation were solved.

$$(hT)_t + (uhT)_x + (vhT)_y = (hD_x T_x)_x + (hD_y T_y)_y + S \qquad ......4.1$$

where, T = rise in salinity (ppt) or temperature, u,v =horizontal velocity components in x, y directions (m/s), D_x , D_y =dispersion coefficients in x, y directions (m^2/s), S = source/sink discharge ($m^3/s/m^2$). This equation is numerically solved by an explicit finite difference scheme.

As it is well known, tow dimensional models are depth averaged and vertically integrated. It can be modeled for horizontal as well as vertical of water bodies using three dimensional models. It is to be noted that model simulates the far field advection-dispersion of the salinity and temperature dispersion in the water ecosystems.

4.5.1 Hydrodynamic Mixing Zone Analysis (CORMIX)

The Cornell Mixing zone Expert System, subsystem 1, 2 and 3 are a series of software elements for the analysis and design of conventional or toxic single port and multiport submerged buoyant, non buoyant or negatively buoyant pollutant discharges into unstratified or stratified watercourse, with the emphasis on the geometry and dilution properties of the initial mixing zone

A typical example of concentration modeling or dispersion modeling of a location is presented in Fig. 4.1 This Expert System is constructed and designed as an analysis tool for discharges and regulators. The Expert System CORMIX1, CORMIX2 and CORMIX3 are applicable to the prediction of mixing behavior of single port and multiport geometry, the characteristics of the Legal Mixing Zone (LMZ) and the zone of toxic dilution. The CORMIX1, CORMIX2 and CORMIX3 collected all input data, conducted hydrodynamic analysis, summarize dilution characteristics including design changes in order to examine dilution characteristics. This chapter is discussed about the modeling application in various ecosystems.

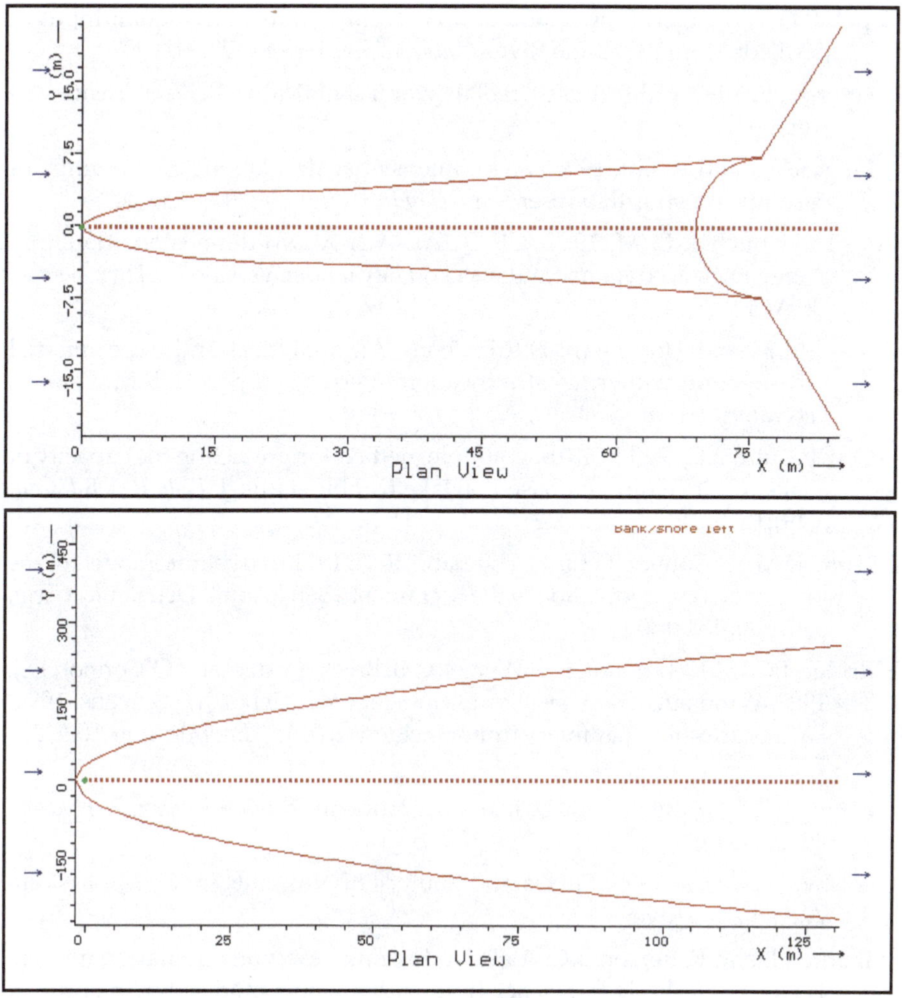

Fig. 4.1 Typical Concentration Modeling of Salinity Dispersion in a Water Body

References

Bloss, S. and Harleman, D.R.F., 1980. Effects of wind induced mixing on the seasonal thermocline in lakes and reservoirs. In:I.T. Carstens and T. McClimans (Editors), *2nd International Symposium on Stratified Flows*, Vol. I, Trondheim, Norway.

Canfield, D.E.JR., R.W. Bachman, 1981: Prediction of total phosphorus concentration, chlorophyll a,and Secchi depths in natural and artificial lakes. *Can. J. Fish. Aquat. Sci.* 38: 414-423.

Canfield and Hoyer. 1988. Regional geology and the chemical and trophic state characteristics of Florida lakes.

Capel, P.D., W. Giger, P. Reichert, and O. Wanner. 1988. Accidental input of pesticides into the Rhine River. *Environ. Sci. Technol.* 22:992–997.

Chatterjee, P. 1991. California suffers its worst chemical spill. *New Scientist.* 10 Aug., p. 12.

Chung, S.W., and R. Gu. 1998. Two-dimensional simulations of contaminant currents in a stratified reservoir. *J. Hydrol. Eng.* 124:704–711.

Cole T.M., Buchak E. M. 1994 (CE-QUAL-W2) A two dimensional laterally averaged, hydrodynamic and water quality model, Version 2.0. *User Manual.* US ACE.

Dake, J.M.K. and Harleman, D.R.F., 1966. An analytical and experimental investigation of thermal stratification in lakes and ponds. R.M. Parsons *Laboratory Technical Report*, 99.

Dillon, P.J. and F.H. Rigler. 1975. A simple method for predicting the capacity of a lake for development based on lake trophic status. *J. Fish. Res. Bd. Can.* 3(19):1519-31.

Di Toro, D.M., O'Connor, D.J. and Thomann, R.V., 1971. A dynamic model for the phytoplankton population in the Sacramento San Joaquin Delta. *Adv. Chem. Ser.*, 106: 131-180.

Eldridge, J.C., M.K. Tennant, L.T. Wetzel, C.B. Breckenridge, and O'Connor, D.J. 1983. Wind effects on gas–liquid transfer coefficients. J.T. Stevens. 1994. Factors affecting mammary tumor incidence, (In): *J. Environ. Eng.* 109:731–752.

Forsberg, C., 1987. Evaluation of lake restoration in Sweden.Scheiz. *Z. Hydrol.*, 49: 260-274.

Goolsby, D.A. et al. 1993. Fluxes and Sources of Nutrients in the Mississioni-Atchafalaya River

Imberger, J. and Patterson, J.C., 1981. A dynamic reservoir simulation model - DYRESM: 5. In: H.B. Fischer (Editor),Transport models for inland and coastal waters. Academic Press, New York, pp. 310-361.

lmberger, J., Patterson, J.C., Hebbert, B. and Loh, I., 1978. Dynamics of reservoirs of medium size. *J. Hydr. Div.*, ASCE, 104: 725-743.

Imberger, J. and Patterson, J.C., 1990. Physical limnology. Adv. Appl. Mech., 27: 303-475.

Jcrgensen, S.E., 1992. Integration of ecosystem theories: A pattern. Kluwer Academic Publ., The Netherlands, 383 pp.

Jcrgensen, S.E., 1983. Eutrophication models of lakes. In: S.E. Jecgensen (Editor), Application of ecological modelling in environmental management, part A. Elsevier Sci. Publ., Ams-terdam, pp. 227-282.

Jcrgensen, S.E., Nielsen, S.N. and Jcrgensen, L.A., 1991. Handbook of ecological parameters and ecotoxicology. Elsevier, Amsterdam, 1263 pp.

Kmet, T., Stra~kraba, M. and Mauersberger, P., 1993. A mechanistic model of the adaptation of phytoplankton photosynthesis. *Bull. Mathemat. Biol.*, 55(2): 259-275.

Martin, J.L., Nirchardson,S.C. McCutcheon, and J. Paul. 1988. *Water Res. Bull.*

Matsuoka, Y., T. Goga and Naito, M., 1986. An eutrophication model of Lake Kasumiganra. *Ecol. Modelling*, 31:201-219.

OECD (1982) Eutrophication of waters: monitoring assessment and control. rganisation for Economic Co-operation and Development, Paris

Orlob, G.T., 1983. One dimensional models for simulation of water quality in lakes and reservoirs. In: G.T. Orlob and G.T. John (Editors), Mathematical Modeling of Water Quality: Streams, Lakes and Reservoirs. *International Series on Applied Systems Analysis*. John Wiley and Sons, pp. 227-273.

Orlob, G.T. and Selna, L., 1970. Temperature variations in deep reservoirs. *J. Hydr. Div.*, ASCE., 96(HY2): Paper 7063.

Patterson, J.C., D.P. Hamilton and Fen'is, J.M., 1994. Modelling of algal growth in the mixed layer of lakes and reservoirs. *Aust. J. Mar. Freshwat. Res.*, 45: 829-845.

Scavia, D., 1980. An ecological model of lake Ontario. *Ecol. Modelling*, 8: 49-78.

Stefan, H. and l~'ord, D.E., 1975. Temperature dynamics in dimictic lakes. *J. Hydr. Div.*, ASCE., Paper 11058:97-114.

Riley, M. J., and Stefan, H. G. (1987). "Dynamic lake water quality simulation model MINLAKE," Report 263, St. Anthony Falls Hydraulic Laboratory, University of Minnesota, Minneapolis, MN.

Riley, M.J. and Stefan, H.G., 1988. MINLAKE: A dynamic lake water quality simulation model. Ecol. Modelling, 43: 155-182. with field observations. *Ecol. Modelling*, 31: 185-199.

USCE, 1986. CE QUAL RI: A numerical one-dimensional model of reservoir water quality, user's manual. Environmental andwater quality operational studies instruction report E-82-1.Department of the Army, U.S. Corps Engineers, Washington,DC, 427 pp.

Vincent, W.F., Gibbs, M.M. and Spigel, R.H., 1991. Eutrophication processes regulated by a plunging river inflow. *Hydrobiologia*, 226: 51-63.

Watanabe, M., Harleman, D.R.F. and Vasiliev, O.F., 1983. Two and three-dimensional mathematical models for lakes and reservoirs. In: G.T. Orlob and G.T. John (Editors), Mathematical Modeling of Water Quality: Streams, Lakes and Reservoirs. International Series on Applied Systems Analysis. John Wiley and Sons, pp. 274-336.

□□□

Chapter 5

Modeling: Methodologies

5.1 Introduction

The water quality modeling methodologies are started from conceptual models such as defining the problem in a particular region of interest like fresh water ecosystem, coastal ecosystems, ocean ecosystems, the following methodology can be studied carefully for water quality modeling.

- Identification of problems in the region of interest
- Monitoring of field data using standard methods like APHA, USGS, BIS etc.
- Evaluation of primary data for the toxins, higher amount of concentration from the ambient conditions
- Processing of primary data to secondary data
- Based on the primary data and secondary data, suitable modeling software can be used. If the temperature of the region is higher than the ambient, it should be used Qual2E, MIKE21 etc.
- Modeling results can be compared with field data for calibration and validation.

Water quality modeling methodological can be divided into following steps.

- Defingin model area
- Collection of field data
- Modeling Assumption
- Calibration and validation

5.2 Defining Model Area

The model area can be defined using previous experiences. hydrodynamic modeling deals with water surface elevation, velocity, current, speed and direction etc. water quality modeling deals with temperature plume, dissolved oxygen concentration, *etc*. The model area for hydrodynamic modeling can be considered with bay, lagoon, mud flats, streams, rivers *etc*. in water quality modeling considered up to the plume getting into ambient condition.

5.3 Collection of Field Data

The field data can be collected from the region with the help of hydrographic surveyor and water quality experts. The physical parameter like currents, tides, depths contour, dye dispersion studies, discharge amounts, sediment size, pH, solids, light transparency, is to be collected from the study areas.

The physical and chemical properties of industrial effluent are characteristic of the climatic, geochemical, geomorphological and pollution conditions prevailing in the drainage basin and the underlying aquifer and industrial processes. The biota in the surface water is governed entirely by various environmental conditions that determine the selection of species as the physiological performance of the individual organisms. The primary production of organic matter, in the form of phytoplankton and macrophytes is more intense in lakes and reservoirs than in rivers. In contrast to the chemical quality of water bodies, which can be measured by suitable analytical methods, biological quality is a combination of both qualitative and quantitative characterisation. This can be carried out in two levels:

- Response of individual species to changes in its environment.
- Response of biological communities to changes in their environment.

The sample collected should be small in volume, enough to accurately represent the whole water body. The water sample tends to modify itself to the new environment. It is necessary to ensure that no significant changes occur in the sample and preserve its integrity till analysed. The essential objectives of water quality assessment are to:

1. Define the status and trends in water quality of a given water body.
2. Analyse the causes for the observed conditions and trends.
3. Identify the area specific problems of water quality and provide assessments in the form of management to evaluate alternatives that help in decision-making.

Sampling sites for the industrial effluents are selected to represent the water quality at different points and depths. Generally three sampling sites are selected for monitoring.

Inlet: the point where the principal feeder opens into the industries.

Center: the point that gives the general water quality of the industrial water pipelines.

Outlet: the place where the outlet occurs to discharge into any water body.

Generally three types of sampling are adopted for collecting water samples.

Grab or Catch sampling: the sample is collected at a particular time and place that represents the composition of the source at that particular point and time.

Composite sampling: a mixture of grab samples is collected at the same sampling point at different time intervals.

Integrated sampling: a mixture of grab samples collected at different points simultaneously.

The quality of water varies with time in a water body due to various natural and human induced factors. The monitoring has to be done in a way that records all the changes in the quality. The sampling frequencies generally adopted in monitoring are:

1. Weekly sampling for one year.
2. Consecutive day sampling during study period.
3. Hourly sampling for 24 hours.
4. Monthly sampling at predefined time.
5. Seasonal sampling at predefined time
6. Yearly integrated sampling at predefined time
7. Continuous sampling up to more than a year

Variations in water quality are mainly due to changes in the concentrations of the components of the water flowing into the water body. These variations can be man-made or natural and can either be cyclic or random.

Random *variations*: due to spasmodic, often unpredictable events such as accidental oil spills, sewage leaks, overflows, *etc.*

Cyclic *variations:* may be a result of regular seasonal changes triggering certain natural processes such as rainfall, snowmelts and seasonal temperature changes, altering the ecosystem. Seasonal growth and decay of vegetation will also rise due to cyclic changes in the composition of water.

In lakes, the mass of water and good lateral mixing provide inertia against any rapid modifications due to inputs and outputs.

The sampling container should not react with the sample, be of adequate capacity to store the sample and be free from contamination. The following table 5.1 shows the different container for various parameters.

Table 5.1 Sampling Containers

Parameters	Containers
BOD, DO,	BOD bottles
Inorganic	Plastic bottles
Heavy metals	Plastic bottles with out reacted by Heavy metals
Other parameters	Preferably plastic cans

Grab sampling was done at the inlet, center and outlet in most of the water bodies studied to assess their physical and chemical qualities at monthly intervals, expect during some seasons when the center of the lake is not accessible due to excessive growth of water hyacinth.

The samples are collected in thoroughly cleaned 2.5-litre inert plastic containers, which are rinsed with distilled and lake, river, reservoir, coastal and ocean water before collection.

Water samples are collected in a sampling bottle avoiding floating materials. The stoppers of the sample containers are closed properly to prevent outside contamination by immersed in the water body. The containeris labelled describing the name of the water body, date, time, sampling point, and conditions under which it was sampled.

Between the time a sample is collected and analysed in the laboratory, physical, chemical and biochemical reactions may take place in the sample container leading to changes in the intrinsic quality of the sample, making it necessary to prevent or minimize these changes with suitable preservatives such as alcohol and mercuric chloride. Highly unstable parameters such as pH, temperature, transparency, free carbon di oxide, dissolved oxygen, etc. are measured at the sampling site. The preservation procedure includes keeping the samples in the dark, adding chemical preservative, lowering the temperature to retard reactions, or combinations of these. The preservation methodologies are showed in table 5.2.

Table 5.2 Preservation Methods for Various Parameters

Experiment	**Preservative**	**Max. holding time**
BOD	Cool, 4°C	4 hours
Calcium	Cool, 4°C	7 days
Chloride	Cool, 4°C	7 days
COD	Cool, 4°C	24 hours
Dissolved Oxygen*	Fix on site	6 hours
Fluoride	Cool, 4°C	7 days
Magnesium	Cool, 4°C	7 days
Nitrate + Nitrite	Cool, 4°C	24 hours
PH	None	6 hours
Phosphorus*		
Dissolved	Filter on site using 0.45μm filter	24 hours
Inorganic	Cool, 4°C	24 hours

Contd...

Table Contd...

Ortho	Cool, 4°C	24 hours
Total	Cool, 4°C	1 month
Potassium	Cool, 4°C	7 days
Specific conductance	Cool, 4°C	24 hours
Sodium	Cool, 4°C	7 days
Heavy metals		
Cadmium	2 ml conc. Nitric acid/L sample	6 months
Chromium	2 ml conc. Nitric acid/L sample	6 months
Copper	2 ml conc. Nitric acid/L sample	6 months
Iron	2 ml conc. Nitric acid/L sample	6 months
Lead	2 ml conc. Nitric acid/L sample	6 months
Nickel	2 ml conc. Nitric acid/L sample	6 months
Zinc	2 ml conc. Nitric acid/L sample	6 months

(Source: Analytical Methods Manual, Water quality branch, Environment Canada, 1981)

The parameters analysed to assess the water quality are broadly divided into:

Physical parameters: Colour, Temperature, Transparency, Turbidity and Odour.

Chemical parameters: pH, Electrical Conductivity (E.C), Total Solids (TS), Total Dissolved Solids (TDS), Total Suspended Solids (TSS), Total Hardness, Calcium Hardness, Magnesium Hardness, Nitrates, Phosphates, Sulphates, Chlorides, Dissolved Oxygen (D.O), Biological Oxygen Demand (BOD), Chemical Oxygen Demand (COD), Fluorides, Free Carbon-di-oxide, Potassium and Sodium.

Heavy metals: Lead, Copper, Nickel, Iron, Chromium, Cadmium and Zinc.

Biological parameters: The biological parameters involved the qualitative analyses of planktons (zooplankton and phytoplankton).

5.4 Calibrations and Validations

Since prediction is the most important component of water quality modeling study, accuracy of the predictive models and hence the uncertainties involved in the predicted results must be known before hand to make any meaningful judgement with regards to the impacts of the proposed project. Uncertainty is an inevitable component of all predictions. The extent of uncertainty that invariably creeps into prediction depends upon the quality of data and the nature of model used. Information on impacts based on predictions is one of the most scientific and objective components of EIA study (Peter 1998). The extent of decision maker being wrong or right with regards to the implementation of the proposed activity is greatly influenced by the accuracy of the model prediction in addition to many other subjective elements included in EIA. Because of its sensitive nature and ability to influence the final results of EIA, prediction of impacts has been recognised as one of the major sources of uncertainty in water quality modeling studies. Since it is an accepted truth that, uncertainty is an inherent and inevitable component of all predictions, it must be minimised to a level that is acceptable from decision-makers point of view.

Since Mathematical models are extensively used in water quality modeling studies for the prediction of the quality of the major components of physical, chemical and biological nature and also in the evaluation of various pollution control alternatives, these models must be calibrated and validated in order to minimise errors involved in the prediction. The usefulness of an appropriately calibrated model in evaluating water pollution control strategies is demonstrated in one of the case studies of water quality modeling a River and lake (Seng Lung *et al.,* 1995).

Identification of sensitive model parameters through sensitivity analysis and their estimation through laboratory and field studies, subsequent calibration and validation of the one dimensional river water quality model under steady state condition were the main objectives of the study. Performance evaluation of the calibrated model was also a part of the study.

Model calibration is the first stage testing or tuning of the model to a set of field data not used in the original construction of the model. Such tuning is to include consistent and rational set of theoretically defensible parameters and inputs (Thomann 1982). Model calibration is actually the process by which one obtains estimates for the model parameters through the comparison of field observations and model predictions. Even if the steady state condition is assumed, the environmental parameters can still vary due to random changes of temp, stream discharge, time of day, and general weather conditions. Due to this inherent dynamic nature of the environment, discrepancies between the predicted and observed results are bound to occur. How credible the model is, or what is the level of confidence that can be placed on the model predictions? Is all depending on the range of discrepancies mentioned above? Such discrepancies must be minimised to the extent possible by identifying and minimising sources of error ie measurement errors, conceptual error in the model, approximation errors due to nature of model being calibrated.

The effect of measurement errors can be minimised by optimising data collection procedures like, collecting data in most sensitive locations and by collecting optimum number of replicates. A first order error analysis or sensitivity analysis can be used to identify critical measurements and sampling locations. It is also desirable to estimate the variance of measurements between replicate samples at select locations during study.

It is the testing of the calibrated model against the additional set of field data preferably under different environmental conditions (river flow, waste load *etc.*), to further examine the range of validity of the calibrated model. Collection of data for validation is such that, calibration parameters are fully independent of the validation data. The model so verified can be used for forecasting of water quality under a variety of perturbed environmental conditions. However, it is apparently rare that, following a forecast and subsequent implementation of environmental control program, an analysis is made of the actual ability of the model to predict water quality responses, such an exercise is regarded as "post audit." of the model. It is nothing but the subsequent examination and verification of the models predictive ability following the implementation of environmental control program.

The same model that is used herein for the present study was also used in the prediction of water quality during water quality modeling studies conducted earlier, for the same river but with the assumed waste loads of the proposed industry.

One-dimensional steady state river water quality model, Qual-2E version developed by US –EPA (Brown *et al.*, 1987) is selected for the present study because, the same model is employed in the earlier EIA studies at the same site. Since it is observed that, water quality parameters in the stream vary predominantly in the longitudinal direction, one-dimensional approximation was assumed. The relevant processes considered in the present study can be simulated with this model. The model is one of the widely accepted water quality simulation tools for waste load allocation purpose and water quality impact analysis studies (William *et al.*, 1992).

Sensitivity analysis is performed to identify the critical model parameters and sensitive sampling locations. Headwater BOD is found to be the most sensitive parameter followed by de-oxygenation and reaeration rate. Dispersion coefficient is found to be least sensitive parameter of the model. This parameter is seems to have been included only to take care of numerical dispersion.

River flows, velocity, and cross sectional details at different locations of the river are to be collected. Since the velocity of a river varies with width and depth due to functional effects, the mean velocity is estimated over 0.8 and 0.2 times the total depths of water in the stream. Average of the two velocities is considered as the representative velocity. However, in shallow rivers, even a single velocity of 0.6H would be sufficient. Flow is obtained by the estimation of a velocity over the reach of the river using dye. The time of travel (t) between the locations of the river is estimated using the tracer (Rodamine).

The average velocity of the river was thus found to be 0.05m / Sec. Since the river cross section closely resembled trapezoidal shape, assumption of trapezoidal cross section was considered appropriate. Bottom width of the river varied from 15 to 50 mt., whereas the top width varied from 20 to 60 mt. Side slopes varied from 0.015 to 0.1 and 0.03 to 0.1 for both sides of the canal. Longitudinal bed slope of the canal varied from 6.6E-07 to 1.3E-06.

Calibration process is mainly based on the field measurements that helps in choosing the empirical coefficients in water quality models and also in the verification of the consistency of the model's initial and boundary condition with that of the in-stream measurements. Calibration of the hydrodynamic part of the model was first carried out by comparing simulated hydrodynamic variables (depth and velocity) with the measured ones. Next, the calibration of the process compartment of the model was carried out sequentially by using transformation kinetic parameters. The order of calibration is;

temperature, BOD, DO, nitrification and phosphorous related parameters respectively.

While calibrating water quality process part of the model, estimated parameters are fixed, few parameters are extracted from standard modeling literature, remaining parameters are obtained by tuning them till the observed and predicted results closely matched.

The calibrated model is used to predict the water quality with an independent set of data as a part of validation exercise. Results of model predictions are fairly good, and performance of the model is further confirmed through statistical evaluation of the results. Statistical analysis of the model results indicate that, simulated CBOD results are relatively in best agreement with the measured values with coefficient of regression 0.9, coefficient of variation 7.2%, RMS 19.2% and relative error 8%. One dimensional steady state water quality model is calibrated and validated. Model coefficients are estimated through field and laboratory studies. Performance of the calibrated model was evaluated through statistical techniques. Model output is found to be very sensitive to headwater quality. Sensitivity of the Carbonaceous Biochemical Oxygen Demand decay rate is found to be increasing towards downstream of the outfall. Model coefficients related to settling rates of CBOD, Nitrogen and Phosphorous fraction are found to be moderately sensitive. Among all the model coefficients of the Qual-2e, dispersion constant is found to be least sensitive and seems to have been included only to take care of numerical dispersion. Based on statistical evaluation of the calibrated model it is evident that, the model has been calibrated reasonably well. Among the state variables presented here, prediction of CBOD results found to be better with relative error of 8%, considering the fact that 20 % deviation is normally acceptable in standard river water quality modeling practices.

References

Brown, L.C., and Barnwell, T.O., Jr. 1987. " The Enhanced Stream water quality Models, Qual-2E and Qual-2E UNCAS: *Documentation and Users Manual.*" EPA /600/3-87/007, Envir. Res. Lab., Envir. Protection Agency (EPA), Athens, Ga.

Peter., W. 1998. Environmental Impact Assessment: Theory &Practice. Academic Division of Unwin Hyman Ltd., 62-97.

Seng Lung, W.U., and Larson C.E. 1995. " Water Quality Modelling of Upper Mississippi River and Lake Peppin. *Journal of Environmental Engineering*, Vol.121, No.EE10, 691-699

Thomann., R.V., and Muller., J.A. 1982 " Verification of Water Quality Models." *Journal of Environmental Engineering*, Vol.108, No.EE 5, 923-940.

William, B.M., George, L. B., Thomas, M.G., and Kay, M. 1992. Hand Book, "Stream Sampling for Waste Load Allocation Application." US EPA, Office of Research and Development, Washington, D.C 20460.

999

Chapter 6

Modeling: Rivers and Streams

6.1. Introduction

A river is a large natural waterway. The source of a river may be a lake, a spring, or a collection of small streams, known as headwaters. From their source, all rivers flow downhill, typically terminating in the ocean. The mouth, or lower end, of a river is known as its base level.

A river water is confined to a channel, made up of a stream bed between banks. Most rainfall on land passes through a river on its way to the ocean. Smaller side streams that join a river are tributaries. The scientific term for any flowing natural waterway is a stream; so in technical language, the term river is just a shorthand way to refer to a large stream. Rivers throughout history have been the source and support for civilization, and many major cities today are near a river of some sort (Fig. 6.1).

6.1.1 Topography

A river conducts water by constantly flowing from a higher to a lower elevation, thereby converting the potential energy of the water into kinetic energy. It starts to form loops and snakes through the plain by eroding the river banks. Sometimes the river will cut off a loop, shortening the channel and forming an oxbow lake from the cut off section. Rivers that carry large amounts of sediment may develop conspicuous deltas at their mouths, if conditions permit. Rivers, whose mouths are in saline tidal waters, may form estuaries. River mouths may also be rias.

Fig. 6.1: A typical river ecosystem

Over a time, rivers cut away at their beds especially in the steeper headwater regions, eventually forming a more gentle gradient. Although the following classes are a useful simplified way to visualize rivers, it is important to recognize there are other factors at work here. Gradient is controlled largely by tectonics, but discharge is controlled largely by climate and sediment load is controlled by various factors including climate, geology in the headwaters, and the stream gradient.

- Youthful river - a river with a steep gradient that has very few tributaries and flows quickly. Its channels erode deeper rather than wider. (*Eg*: Brazos River, Trinity River, Ebro River)

- Mature river - a river with a gradient that is less steep than those of youthful rivers and flows more slowly than youthful rivers. A mature river is fed by many tributaries and has more discharge than a youthful river. Its channels erode wider rather than deeper. (*Eg*: Mississippi River, Ohio River, Thames River)

- Old river - a river with a low gradient and low erosive energy. Old rivers are characterized by flood plains. (*Eg*: Tigris River, Euphrates River, Indus River)

- Rejuvenated river - a river with a gradient that is raised by the earth's movement.

Where a river descends quickly over sloped topography, rapids with waterfalls occur. Rapids are often used for recreational purposes Waterfalls are sometimes used as sources of energy, via watermills and hydroelectric power plants.

Rivers begin at their source in higher ground, either rising from a spring, forming from glacial meltwater, flowing from a body of water such as a lake, or simply from damp, boggy places where the soil is waterlogged. They end at their base level where they flow into a larger body of water, the sea, a lake, or as a tributary to another river. In arid areas rivers sometimes end by losing water to evaporation and percolation into dry, porous material such as sand, soil, or pervious rock. The area drained by a river and its tributaries is called its watershed, catchment basin or drainage basin.

Starting at the mouth of the river and following it upstream as it branches again and again, the resulting river network forms a dendritic structure that is an example of a natural random fractal. Each river has its own unique river regime. This describes the general discharge of the river throughout the course of the year.

6.1.2 Geopolitical boundaries

Rivers have been important historically in determining political boundaries. For example, the Danube is a longstanding border of the Roman Empire, and today forms most of the border between Bulgaria and Romania. The Mississippi in North America, and the Rhine in Europe, are major east-west boundaries in those continents. The Orange River forms the boundary between various provinces and countries along its route in Africa.

Rating systems

- *International Scale of River Difficulty* : The scale is used to rate the challenges of navigation particularly those with rapids. Class I is the easiest and Class VI is the hardest.
- *Strahler Stream Order* : The Strahler Stream Order is a method to rank rivers based on the connectivity and hierarchy of contributing tributaries. Headwaters are first order while the

Amazon River is twelfth order. Approximately 80 percent of the rivers and streams on Earth are of the first and second order.

6.1.3 Biology

The flora and fauna of rivers are much different from those of the ocean because the water is fresh (non-salty). The following figures (Fig. 6.2) are aquatic organisims which are living in the river ecosystems.

6.2a. Blue green algae.

6.2b : diatoms

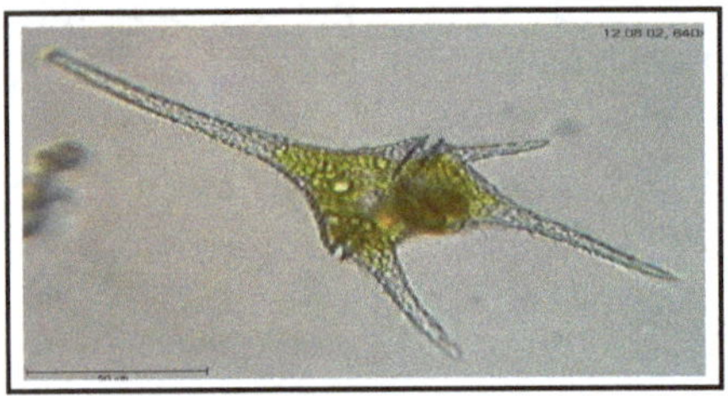

6.2c. Ceratium

6.2d. Meganctiphanes

Fig. 6.2. Aquatic Organisms in River Ecosystem

6.1.4 Flooding

Flooding is a natural part of a river's cycles (Nezette Rydell, 1997). Human activity, however, has upset the natural way flooding occurs by walling off rivers and straightening their courses. Removal of bogs, swamps and other habitat. Today, many believe that the rebuilding of salmon runs is contingent upon reproducing the same environment. In wintertime, rivers can be apart, it can lead to flash flooding downstream.

6.1.5 Flow

A common misconception, particularly amongst schoolchildren and college students in North America (Fermat,1997), is that most, or even all, rivers flow from north to south. Rivers in fact flow downhill.

Sometimes downhill is from north to south, but equally it can be from south to north, and usually is a complex meandering path involving all directions of the compass (William C. Philips, 1991). Studying the flows of rivers is one aspect of hydrology (Matt Rosenberg, 2006).

6.1.6 Management

In its natural state a river can be managed or controlled to make them more useful and less disruptive to human activity.

- Dams or weirs may be built to control the flow, store water, or extract energy.
- Levees may be built to prevent run-off of excess river water in times of flood.
- Canals connect rivers to one another for water transfer or navigation.
- River courses may be modified to improve navigation, or straightened to increase the flow rate.

River management is an ongoing activity as rivers tend to 'undo' the modifications made by man. Dredged channels silt up, sluice mechanisms deteriorate with age, levees and dams may suffer seepage or catastrophic failure.

6.1.7 World Longest Rivers

It is difficult to measure the length of a river, the more precise the measurement, Also, it is difficult to determine where a river begins or ends, as very often, upstream rivers are formed by seasonal streams, swamps, or changing lakes.

These are average measurements.

1. Nile (6,690 km)
2. Amazon (6,452 km)
3. Mississippi-Missouri (6,270 km)
4. Yangtze (Chang Jiang) (6,245 km) (Simon Singh, 1997)
5. Yenisey-Angara (5,550 km)
6. Huang He (Yellow) (5,464 km)
7. Ob-Irtysh (5,410 km)

8. Amur (4,410 km)
9. Congo (4,380 km)
10. Lena (4,260 km)

6.2 Fresh Water Resources in India

Fresh water as a commodity generates concern being an exhaustible resource and due to the environmental issues related to its degradation. With a phenomenal development of water resources since independence, India has successfully met water requirements for different designated usages. Preserving the quality and Quantity of freshwater resources however, is becoming the most pressing of many environmental challenges on the national horizon. Perhaps, because water is considered a cheap readily available resource, people fail to aware just how much stress human demands for water are placing on natural ecosystems.

India receives an average annual rainfall equivalent of about 4,000 billion cubic metres (BCM). This only source of water is unevenly distributed both spatially as well as temporally. Most of the rainfall is confined to the monsoon season, from June to September, and levels of precipitation vary from 100 mm a year in western Rajasthan to over 9,000 mm a year in the northeastern state of Meghalaya. With 3,000 BCM of rainfall concentrated over the four monsoon months and the other 1,000 BCM spread over the remaining eight months, our rivers carry 90% of the water during the period from June-November every year. Thus, only 10% of the river flow is available during the other six months in a year. National level statistics for water availability mask huge disparities from basin-to-basin and region to region. Spatially, the utilizable resource availability in the country varies from 18,417 cubic meters in the Brahmaputra valley to as low as 180 cubic metres in the Sabarmati basin (Chitale 1992). Even within the Ganga basin, water availability varies from 740 cubic meters in Yamuna to 3,379 cubic meters in the Gandak. Rajasthan, for instance, with 8% of the country's population has only 1% of the country's water resources while Bihar with 10% of population has just 5% of the water resources. Thus, while India is considered rich in terms of annual rainfall and total water resources, its uneven geographical distribution causes severe regional and temporal shortages of water resources.

Of 4,000 BCM of available water from rainfall , the mean flow in the country's rivers is about 1,900 BCM. Out of this, only 690 BCM is utilisable. With 177 BCM of live storage created by the existing major and medium dam projects and another 75 BCM of storage from projects under construction, there is still a gap of 440 BCM of water, which is utilised (CWC 1997). The latest assessment of replenishable ground water resources has been made at 431.9 BCM by the Central Ground Water Board through a large volume of hydrologic and related data. This is the total of poten-tial due to natural recharge from rainfall and due to recharge contributions from canal irrigation. The utilisable ground water re-sources have been assessed at 395.6 BCM (70.0 BCM for domestic and industrial uses and 325.6 BCM for irrigation). The CGWB has also assessed the quantum of static ground water resources at 10,812 BCM.

Water availability from other sources and through desalinisation of sea is considered negligible in view of the high cost. The assessed gross available and utilisable water resources of the country, based on conventional tech-nology, are therefore 2,384 BCM and 1,086 BCM, respectively. With an estimated population of one billion in 200, the available and utilisable water resources per capita per year are 2,384 cubic meter and 1,086 cubic meter respectively against an estimated availability of 6008 cubic meter in 1947. This itself, gives a broad indication of the growing resource scarcity in India in the fifty eight years since independence.

Since independence, India has witnessed an unprecedented increase in population. From a population of about 343 million in 1947, the population has grown at a rate of 2.04% to cross the 1,000 million mark in 2001. With an increasing number of mouths to feed, there has been an additional pressure on agriculture resulting in an increase in net sown area from 119 million hectares in 1951 to 142 million hectares in 1997; high cropping intensity has also resulted in an increased demand for water resources. Domestic water need in the urban areas has also grown notably with the current urban population at 4.5 times the population level in 1950s (TERI 1998). The water re-quirement of the manufacturing sector has increased in proportion to the increase in the sector's share in GDP from about 12% in 1950s to 20% in 1990s. Further, there is a substantial variance in the different user sectors—agriculture, domestic and industry, vis-à-vis their share of water demands, resource pricing structure and usage efficiencies, which creates intersectoral com-petitions and conflicts. The agriculture sector, for instance, accounts for about 95% of the total water demand with the subsidised and free regime of supply of power and water resulting in the over-exploitation and inefficient usage of water. The high resource cost for industries, on the other hand, cross- subsidises the water consumed by the other sectors (TERI 2001).

The demand for fresh water has been identified, as the quantity of water required to be supplied for specific use and includes consumptive as well as necessary non-consumptive water requirements for the user sector. The total water withdrawal/utilisation for all uses in 1990 was about 518 BCM or 609 m 3 /capita/year. Estimates for total national level water requirements, through an iterative and building block approach, have been made for the years 2010, 2025 and 2050 based on a 4.5% growth in expenditure and median variant population

Although about 88% of the population, on average, has been covered with organized drinking water supply in class I cities and class II towns of the country, there is a huge disparity in quantity of water supplied. Of the 299 class I cities, only 77 cities have cent percent water supply coverage. The per capita water supply also ranges from as low as 9 Litres Per Capita per Day (LPCD) in Tuticorin to as high as 584 lpcd in Triuvannamalai (CPCB 2000a). Similarly, 203 of the 345 class II towns have low per capita supplies of less than 100 lpcd (CPCB 2000b). Table 11.3 highlights the wide disparity that exists in the per capita projections of the United Nations. The country's total water requirement by the year 2050 will become 1,422 BCM, which will be much in excess of the total utilisable average water resources of 1,086 BCM. At the national level, it would be a very difficult task to increase the availability of water for use from the 1990 level of approximately 520 BCM to the desired level of 1,422 BCM by the year 2050 as most of the undeveloped utilisable water resources are concentrated in a few river basins such as the Brahmaputra, Ganga, Godavari, and Mahanadi. The Accelerated Rural Water Supply Pro-gramme (ARWSP) and Rajiv Gandhi National Drinking Water Mission (RGNDWM). The status of coverage of habitations as on 1 April 1999, shows that about 267 347 habitations (18.6% of total) are still either not covered or partially covered. Cent percent rural habitat resources in India. Urban centres contribute more than 25% of the sewage generated in the country. The smaller towns and rural areas do not contribute significant amounts of sewage due to the low per capita water supply. Wastewater generated normally percolates into the soil or evaporates.The CPCB conducted a survey in 1994 - 95 (CPCB 2000a, CPCB 2000b) on water supply and wastewater generation, collection, treatment and disposal in 299 class I cities and 345 class II towns of the country. The survey findings indicated that, most cities did not have organised wastewater collection and treatment facilities. The status of wastewater generation, collection, and treatment in urban areas is as given in Table 11.4. It also emerged that the class I cities and class II towns of Maharashtra, Delhi, Uttar Pradesh, West Bengal, and Gujarat contribute 59% of the total wastewater generated in the country.

The annual precipitation including snowfall, which is the main source of the water in the country is estimated to be of the order of 4000 cu.km. For the purpose or rainfall country has been divided into 35 meteorological sub-divisions. The Resources potential of the country, which occurs as natural run off in the rivers is about 1869 cu.km. as per the basin wise latest estimates of Central Water Commission, considering both surface and ground water as one system. Ganga-Brahmaputra-Meghna system is the major contributor to total water resources potential of the country. Its share is about 60 percent in total water resources potential of the various rivers. Based on 1991 census, the per capita availability of water works out to 220 cubic metre cu.m.). Due to various constraints of topography, uneven distribution of resource over space and time, it has been estimated that only about 1122 cu.km. of total potential of 1869 cu.km. can be put to beneficial use, 690 cu. km. being due to surface water resources. Again about 40 percent of uitlisable surface water resources are presently in Ganga-Brhamputra-Meghna system. In majority of river basins, present utilisation is significantly high and is in the range of 50 percent to 95 percent of utilisable surface resources. But in the rivers such as Narmada and Mahanadi percentage utilisation is quite low. The corresponding values for these basins are 23 percent and 34 percent respectively.

The distribution of water resources potential in the country shows that as against the national per capita annual availability of water as 2208 cu. m. the average availability in Brahmaputra and Barak is as high as 16589 cu m. while it is as low as 360 cu.m. in Sabarmati basin. Brahmaputra and Barak basin with 7.3 percent of geographical area and 4.2 percent of population of the country has 31 percent of the annual water resources. Per capita annual availability for rest of the country excluding Brahmaputra and Barak basin works out to about

1583 cu.m. Any situation of availability of less than 1000 cu. m. per capita is considered by international agencies as scarcity conditions. Cauvery, Pennar, Sabarmati, East flowing rivers and West flowing rivers are some of the basins which fall into this category.

Table 6.1: Basin wise Surface Water Potential of India

S. No.	Name of the River Basin	Average annual potential in river Cubic Km/Year
1.	Indus (up to Border)	73.31
2.	a) Ganga	525.02
	b) BBrahmaputra Barak & Others	585.60
3.	Godavari	110.54
4.	Krishna	78.12
5.	Cauvery	21.36
6.	Pennar	6.32
7.	East Flowing Rivers Between Mahanadi & Pennar	22.52
8.	East Flowing Rivers Between Pennar and Kanyakumari	16.46
9.	Mahanadi	66.88
10.	Brahmani & Baitarni	28.48
11.	Subernarekha	12.37
12.	Sabarmati	3.81
13.	Mahi	11.02
14.	West Flowing Rivers of Kutch, Sabarmati including Luni	15.10
15.	Narmada	45.64
16.	Tapi	14.88
17.	West Flowing Rivers from Tapi to Tadri	87.41
18.	West Flowing Rivers from Tadri to Kanyakumari	113.53
19.	Area of Inland drainage in Rajasthan desert	NEG.
20.	Minor River Basins Drainage into Bangladesh & Burma	31.00
	Total	**1869.35**

6.3 Case study of Krishna River Water Quality Modeling

The rate at which we deplete and degrade our fresh aquatic resources poses a great threat to our future life support system. The rise in human population exploits more of natural resources and this is met through the growth of industries specifically chemicals and petrochemicals, urbanisation, deforestation and intensive agricultural practices. The industries and urban sprawl discharge the waste into river. The deforestation process itself aggravates the sedimenta transport into the streams. And the use of chemicals in the crops for better production contaminates ground water through percolation, and rivers and lakes through surface run-off. All these sporadic degrading activities have lead to gradual deterioration in the quality of surface and subsurface water. The loss of quality is causing health hazards and death of human, livestock and death of aquatic organisms, crop failure and loss of aesthetics. Keeping in view the importance of good water quality, the Central Pollution Control Board (CPCB), in 1976, initiated a series of integrated river basin studies all over the country. CPCB in collaboration with the State Pollution Control Boards (SPCB) established the Water Quality Monitoring (WQM) network in the country. The CPCB has identified river stretches all over the country, which have been polluted to the maximum extent. The Krishna river, which is one such polluted rivers of the country, flows in the states of Maharashtra, Karnataka and Andhra Pradesh. The present study is taken up for the monitoring, identification and suggesting preliminary measures of water pollution control in the Satara-Sangli stretch (stretch-I) of the Krishna basin in Maharashtra with the help of Geographic Information System (GIS). The stretch-I, also known as the country's sugar-belt, has been identified by CPCB and MPCB (Maharashtra Pollution Control Board) for the restoration of water quality under the National River Action Plan (NRAP).

6.3.1 Satara - Sangli Stretch

The Satara-Sangli stretch of the Krishna river in Maharashtra is known as the sugar belt of the country. The Krishna river flows with a southeasterly trend in Maharashtra, traversing a distance of 280 kms from Mahabaleshwar through Satara to Sangli [latitudes 160 00' N - 180 00' N, longitudes 730 30' E - 750 00' E, altitude 150-600 m] on a rocky area. The river is met by river Koyna, about 137 kms away from its source. Numerous seasonal tributaries dissect the basin giving rise to a dendritic pattern of drainage with a medium drainage density.

The total geographical area of the Krishna basin in Satara is 10,816 km2 (4%) and that of Sangli is 8,572 km2 (3.2%). The districts of Satara and Sangli experience a warm humid climate with an annual average precipitation between 600 and 800 mm. Eighty percent of the rainfall in the Krishna basin is influenced by the south west monsoon giving rise to heavy rainfall on the west coast of the western ghats. The mean temperature varies between 22.5 - 25° C. Usually, the geological formations consisting of Deccan traps are rich in basalt and dolomite type rocks. The soils in the basin are *in situ* in nature and a major portion comprises mainly of medium black cotton (b.c) soils. These b.c. soils are the weathered derivations of the Deccan Basalts. Extensive bleaching and weathering of b.c. soils has given rise to red loamy and lateritic soils.

6.3.2 *Water Consumption and Effluent Discharge*

The state of Maharashtra is ranked first in terms of industrial investment in the country. Major industrial sectors are in power, fertiliser, sugar and cement industries. In satara and Sangli fifteen medium to large size sugar industries are located. There are many liquor factories located along the stretch-I. The quantity of water that is consumed for domestic, industrial and irrigation uses are respectively 66, 18 and 3366 MCM. Correspondingly, the amount of effluent that is being discharged from urban, industrial and irrigation are 29, 14 and 673 MCM. From the sugar factories and its surrounding domestic locations about 13400 and 1525 cubic meter of effluents are being discharged everyday.

6.3.3 *A Framework for Monitoring Water Quality in GIS*

River water quality monitoring is the process of regular study of parameters related to river water. It helps determining the quality trend and hence the threshold values for the restoration of water quality to its normal. Different factors those affect the water quality are physical, chemical and socio-economic parameters of the river basin. The present case study is followed up as per this framework. Using GIS, the database on pollution load, the relationship between pollution load with population, fertiliser consumption and factory location, and the river zonation have been assessed and graphically presented. The techniques of river zonation has been reviewed and modified. The prime objectives of using GIS over traditional methods are :

- Effective storage and analysis system for spatial and temporal databases such as maps on geology, geomorphology, soils, land uses and attributes on meteorology, population, water quality etc.,
- Spatial analysis on depicting the source-pollutant relationship,
- Graphical presentations, visual impacts and spatial distribution of graphical outputs on water quality changes, pollution load and relationship with sources and
- Management of river basins by generating buffer zones on the basis of water quality criteria.

6.3.4 Water Quality and Pollution Load at Stretch-I

The stretch-I is about 180 kms. This stretch, covering a total area of 13065.22 km2, is subdivided into three sub-watersheds SW1 (1705.17), SW2 (3545.4) and SW3 (7814.65) km2. A Water Quality Monitoring (WQM) station accompanies each one of these sub-watersheds. The WQMs 1194, 36 and 37 respectively fall within the sub-watersheds SW1, SW2 and SW3. The coverage of Krishna channel within these subwatersheds are respectively 40.92, 300.84 and 531.10 Km2. About 19 water quality parameters, the physical parameters temperature, run-off and turbidity and the chemical parameters pH, hardness, conductivity, alkalinity, DO, BOD, COD, Fcoli, Total Coliform, Nitrogen, Chlorine, Sulphur, Sodium, Calcium, Magnesium and TKN, were studied from their monsoon and non-monsoon readings. While computing the pollution load (Table 3) it was assumed that the river flows in the stretch 365 days a year. The exposure of total population to pollution load in each subwatershed as shown in this table is to correlate their growth trend. Generally along the stretch-I, turbidity and the chemical parameters BOD, COD, Na, Mg, Ca, Cl, TKN and Sulphate show slightly increasing trend over the years (1984-1997) in the downstream direction of river flow. Parameters like pH, N and DO don't show much of variation from the mean. However, the water quality readings of Fcoli and Tcoli are slightly decreasing along the downstream direction. In the individual WQM station the trend in BOD and COD loads, the indicators of organic pollution, show positive and the COD values are quite higher than BOD. The minimum and maximum BOD values during 1997 were 227 and 13241 tonnes year-1 whereas the COD values were 655 and 33453 tonnes year-1. The BOD and COD loads of the stretch, are showing sharp positive trend from 1990 onwards. These indicated that the inflow of pollutants to river has been increasing after 1990. Amongst all the chemical parameters, the load of magnesium was the maximum. The highest Mg-load obtained was 224416 tonnes year-1 in 1988 for SW3. If the load of each pollutant is listed in terms of their total contribution in an year, the sequence in descending order for these pollutants will be Mg, Ca, Na, Sulphate, Cl, N, COD and BOD. First five major pollutants in the sequence are generally from the agricultural sources and the last two are both from both domestic and industrial sources.

6.3.5 Source Identification

GIS is used to organise both spatially and temporally and presenting graphically the pollution load data for each subwatershed over the period 1984 to 1997. For each pollutant the load data for four years is presented which included years of minimum and maximum pollution loads and the pollution loads of starting and ending years. The spatial variation of all the pollutants showed a steady increase in the load towards the downstream direction. This is due to two reasons - (i) the flow rate (cumecs) of river increased in the downstream direction and (ii) the increase in concentration of water quality parameters, though inconsistent, downstream due to addition of wastes from upstream and additional streams. One of the facts for additional increase of concentration downstream is due to increasing number of sugar factories. In 1997, the number of major sugar factories in SW1, SW2 and SW3 were zero, three and five respectively. The effluents those come out of these factories and surrounding urban setup are added to the stream as a fresh input. The waste discharge of these large and medium sugar factories and the surrounding urban setup are in the order of 13400 and 1524 cubic meter per day respectively. The growth or density of population has increased highly along the downstream which has produced such a large quantity of domestic wastes. Between 1951 and 1991, the population growth km-2 in Satara and Sangli are 121 and 141 respectively. The growth of population synchronised with the growth in factories and the consumption rate of fertilisers. The consumption quantities of fertilisers in the agricultural land have increased by more than 3-fold in Satara and 4-fold in Sangli. This explains why the pollution loads from agricultural sources such as Mg, Ca, Na, Sulphate, Cl and N are continuously increasing along the downstream direction.

The spatial relationship between the pollutants (BOD & COD) and the population growth is correlated using three estimators contingency coefficient, Tschuprow's T and Cramer's V. The estimators show good relationships (V=0.67, T=0.56, Contingency Coefficient=0.56) between BOD and population growth and COD and population growth. Therefore, using the overlay techniques the composites BOD - population growth and COD-population growth were produced. The results in composite are classified into good, bad, very bad and worst. For example, good regions have low population growth and low BOD. In a similar way the relationships between the rate of fertiliser consumption with BOD and COD are estimated. The estimators show again good relationship. These analyses supported the fact that the population rise is a dominant factor to increasing pollution load due to domestic and agricultural sources in the downstream direction.

6.3.6 Pollutant Balance

Industrial and domestic wastes contribute to the major rise in BOD and COD concentrations. The total estimated pollution load for Satara and Sangli from agricultural, domestic and industrial (sugar and others) sources are shown. The waste disposals from sugar and distillery factories are the prime sources of BOD and COD loads. The total amount of waste that is being generated, treated and discharged from sugar and other industries in Satara and Sangli districts are mentioned. The balance of pollutants is estimated for 3 subwatersheds. Since the total pollution load for the stretch is coming from SW1 to SW3 through SW2, the share of each subwatershed to pollution load has been computed. In each subwatershed the total addition is estimated. By subtracting the total waste assimilation capacity (WAC) of river from the total addition, the result becomes the net addition of load that will go into the subsequent subwatersheds in the downstream.

6.3.7 Suggestion on River Water Quality Restoration through Zonation

Buffer zones are used in proximity analysis where the distance from either side of river bank is an important criterion in determining suitability or risk. Buffer zones provide storage for floods and pollution control. Buffer strips made of uneven vegetation (grasses, shrubs, trees) attenuate runoff pollutants that would otherwise reach the body of water. The methods of creating buffer zones on both sides of river bank, also known as corridors, are called as river zonation. The present study suggests over an existing river zonation method. The Satara-Sangli stretch is classified into zone A-II by MPCB but the present study found 3 clearly demarcated zones A-I, A-II and A-III contained in SW1, SW2 and SW3 respectively. Each of these subwatersheds is used for buffer zones demarcating the red, orange and green zones. The buffering distance for each zone is considered as per MPCB's distance criteria (MPCB, 1997). The characteristics of watersheds influence the pollution load at sub stretches due to the water quality properties of streamlets those are induced artificially and naturally. For better understanding of pollution loads in a stretch (watershed), it is essential to know the addition of pollutants at sub stretches. In other way saying, the pollution load is better dependent on the sub watershed characteristics than the flow channel.

It is suggested that the river zonation at stretch-I should be done in the sub watersheds rather than along the flow channel. The total areal coverage as per flow channel zonation are 2849 km2 (within 1 km) and 1845 km2 (within 1-2 kms) and the same from the present suggestion of subwatershed zonation are 1653 km2 (within 3, 1, 0.5 kms) and 1897 km2 (within 8, 2, 1 kms). GIS has been utilised in the storage and retrieval of attribute data such as water quality parameters (pollution loads), population density and fertiliser consumption over the spatial database (map) of Satara-Sangli stretch in the Krishna basin. This database is useful in monitoring the trend of pollution load and population growth in the entire watershed between 1984 to 1997. With the aid of map comparison utility in GIS pollution map can be compared with the population, fertiliser and industry location maps. Satara-Sangli stretch of the Krishna river is polluted grossly by the human induced activities in the subwatersheds. The factors for acute pollution of water are:- the intensive use of fertilisers and pesticides in the agricultural land, growth of medium to big size sugar and distillery factories and very high growth of population leading to high domestic load from urban setup. Amongst the physical parameters turbidity values increased and the same results are witnessed after 1990 for chemical parameters such as BOD, COD, Na, Mg, Ca, Cl, TKN and sulphate. For all the pollutants load values increased abruptly for the subwatersheds along the downstream direction. There has been a good relationship between the pollution parameters with the population density.

About 32 lakhs people got exposed to the pollution in 1991. The growth of people synchronised with that of the growth in industries. About 8 major sugar factories are responsible for the most of industrial effluents. Of all sources, the share of agriculture to water consumption and water pollution is the highest. Agricultural sources contributed to 91 % of total waste discharge while the same for domestic and industrial sources were 4.5 % each. It is very much indispensable that some standard economically feasible technologies be adopted to mitigate and reciprocate the process of water quality degradation, and restore the quality back to its normal.

Since prediction is the most important component of the Environmental Impact Assessment study, accuracy of the predictive models and hence the uncertainties involved in the predicted results must be known before hand to make any meaningful judgement with regards to the impacts of the proposed project. Uncertainty is an inevitable component of all predictions. The extent of uncertainty that invariably creeps into prediction depends upon the quality of data and the nature of model used. Information on impacts based on predictions is one of the most scientific and objective components of EIA study [Peter 1998]. The extent of decision maker being wrong or right with regards to the implementation of the proposed activity is greatly influenced by the accuracy of the model prediction in addition to many other subjective elements included in EIA. Because of its sensitive nature and ability to influence the final results of EIA, prediction of impacts has been recognised as one of the major sources of uncertainty in EIA studies. Since it is an accepted truth that, uncertainty is an inherent and inevitable component of all predictions, it must be minimised to a level that is acceptable from decision-makers point of view. Since Mathematical models are extensively used in EIA studies for the prediction of the quality of the major components of environment, and also in the evaluation of various pollution control alternatives, these models must be calibrated and validated in order to minimise errors involved in the prediction. The usefulness of an appropriately calibrated model in evaluating water pollution control strategies is demonstrated in one of the case studies of water quality modelling of upper Mississippi River and lake Peptone, [Seng Lung *et al.*, 1995].

6.3.8 Purpose of the Study

Identification of sensitive model parameters through sensitivity analysis and their estimation through laboratory and field studies, subsequent calibration and validation of the one dimensional river water quality model under steady state condition are the main objectives of the study. Performance evaluation of the calibrated model is also a part of the study.

6.3.9 Calibration

Model calibration is the first stage testing or tuning of the model to a set of field data not used in the original construction of the model. Such tuning is to include consistent and rational set of theoretically defensible parameters and inputs (Thomann., R.V., and Muller., J.A. 1982)). Model calibration is actually the process by which one obtains estimates for the model parameters through the comparison of field observations and model predictions. Even if the steady state condition is assumed, the environmental parameters can still vary due to random changes of temp, stream discharge, time of day, and general weather conditions. Due to this inherent dynamic nature of the environment, discrepancies between the predicted and observed results are bound to occur. How credible the model is, or what is the level of confidence that can be placed on the model predictions? is all depending on the range of discrepancies mentioned above.

Such discrepancies must be minimised to the extent possible by identifying and minimising sources of error (measurement errors, conceptual error in the model, approximation errors due to nature of model being calibrated). The effect of measurement errors can be minimised by optimising data collection procedures like, collecting data in most sensitive locations and by collecting optimum number of replicates. A first order error analysis or sensitivity analysis can be used to identify critical measurements and sampling locations. It is also desirable to estimate the variance of measurements between replicate samples at select locations during study.

6.3.10 Verification and Validation

It is the testing of the calibrated model against the additional set of field data preferably under different environmental conditions (river flow, waste load etc.), to further examine the range of validity of the calibrated model. Collection of data for validation is such that, calibration parameters are fully independent of the validation data. The model so verified can be used for forecasting of water quality under a variety of perturbed environmental conditions. However, it is apparently rare that, following a forecast and subsequent implementation of environmental control program, an analysis is made of the actual ability of the model to predict water quality responses; such an exercise is regarded as "post audit." of the model. It is nothing but the subsequent examination and verification of the models predictive ability following the implementation of environmental control program. The same model that is used herein for the present study was also used in the prediction of water quality during EIA studies conducted earlier, for the same river but with the assumed waste loads of the proposed industry.

6.3.11 Selection of the Model

One-dimensional steady state river water quality model, Qual-2E version developed by US -EPA [Brown *et al.,* 1987] is selected for the present study because, the same model was employed in the earlier EIA studies at the same site. Since it was observed that, water quality parameters in the stream vary predominantly in the longitudinal direction, one-dimensional approximation is assumed. The relevant processes considered in the present study can be simulated with this model. The model is one of the widely accepted water quality simulation tools for waste load allocation purpose and water quality impact analysis studies [William *et al.,* 1992].

6.3.12 Identification of Pertinent State Variable

Sensitivity analysis is performed to identify the critical model parameters and sensitive sampling locations. Headwater BOD is found to be the most sensitive parameter followed by deoxygenation and reaeration rate. Dispersion coefficient is found to be least sensitive parameter of the model. This parameter is seems to have been included only to take care of numerical dispersion.

6.3.13 Hydrodynamic variables

River flows, velocity, and cross sectional details at different locations of the river are collected. Since the velocity of a river varies with width and depth due to functional effects, the mean velocity is estimated over 0.8 and 0.2 times the total depths of water in the stream. Average of the two velocities is considered as the representative velocity. However, in shallow rivers, even a single velocity of 0.6H would be sufficient. Flow is obtained by the estimation of a velocity over the reach of the river using dye. The time of travel (t) between the locations of the river is estimated using the tracer (Rodamine). The average velocity of the river is thus found to be 0.05m / Sec. Since the river cross section closely resembled trapezoidal shape, assumption of trapezoidal cross section is considered appropriate. Bottom width of the river varied from 15 to 50 mt., whereas the top width varied from 20 to 60 mt. Side slopes varies from 0.015 to 0.1 and 0.03 to 0.1 for both sides of the canal (left and right bank w.r.t flow direction). Longitudinal bed slope of the canal varied from 6.6E-07 to 1.3E-06.

6.3.14 Water Quality variables

Water quality data is collected at 12 sampling locations over a stretch of 10.4 Km. Average values of DO, BOD, TKN, NO2 and NO3 and PO4 varied from 9.1 - 12.8, 2-3, 3-8, 0.05-0.1, 1-1.5, and 0.1-0.4 mg/L respectively.

6.3.15 Waste Load

The only wastewater out-fall being discharged into the river is from the National Fertiliser Limited. Average value of wastewater flow is found to be 0.12 m^3 /Sec. The values of BOD, DO, NO2 , NO3, TKN and PO4 are found to be 12, 7.6, 2.2, 2.6, 9 and 2.1 respectively.

6.3.16 Kinetic Coefficients

Based on sensitivity analysis, the nature of wastewater and the process to be simulated. First order error analysis is performed for three locations; upstream and downstream of outfall, and at the location of minimum DO. The results indicated that, sensitivity coefficient of CBOD is relatively more for wastewater flow and CBOD.

6.3.17 Field Estimation of Dispersion Coefficient

Dispersion coefficient is regarded as one of the most important parameters, which characterises the ability of the stream to disperse pollutants. This coefficient combines the effect of diffusion with dispersion and generally known as hydrodynamic dispersion. In the present study, this parameter is estimated through tracer study conducted in the field. Rodamine is used as a tracer based on the traits of an ideal tracer viz., water solubility, ease of detection, minimum background interference, stable, non toxic to human beings, harmless to aquatic life and inexpensive (McCutcheon 1989). Based on field studies, longitudinal dispersion coefficient is estimated to be 1m 2 / Sec and dispersion constant was found to be K= 1223. Refer to Thomann *et al.* (1987) for dispersion coefficient calculations.

6.3.18 Calibration Methodology

Calibration process is mainly based on the field measurements that helps in choosing the empirical coefficients in water quality models and also in the verification of the consistency of the model's initial and boundary condition with that of the in-stream measurements. Calibration of the hydrodynamic part of the model was first carried out by comparing simulated hydrodynamic variables (depth and velocity) with the measured ones.

Next, the calibration of the process compartment of the model was carried out sequentially by using transformation kinetic parameters. The order of calibration is; temperature, BOD, DO, nitrification and phosphorous related parameters respectively. While calibrating water quality process part of the model, estimated parameters are fixed, few parameters are extracted from standard modelling literature, remaining parameters are obtained by tuning them till the observed and predicted results closely matched.

6.3.19 Performance Evaluation of the Model

The calibrated model is used to predict the water quality with an independent set of data as a part of validation exercise. Results of model predictions are fairly good, and performance of the model is further confirmed through statistical evaluation of the results. Statistical analysis of the model results indicate that, simulated CBOD results were relatively in best agreement with the measured values with coefficient of regression 0.9, coefficient of variation 7.2%, RMS 19.2% and relative error 8%.

One dimensional steady state water quality model is calibrated and validated. Model coefficients are estimated through field and laboratory studies. Performance of the calibrated model was evaluated through statistical techniques. Model output is found to be very sensitive to headwater quality. Sensitivity of the Carbonaceous Biochemical Oxygen Demand decay rate is found to be increasing towards downstream of the outfall. Model coefficients related to settling rates of CBOD, Nitrogen and Phosphorous fraction are found to be moderately sensitive. Among all the model coefficients of the Qual-2E, dispersion constant is found to be least sensitive and seems to have been included only to take care of numerical dispersion. Based on statistical evaluation of the calibrated model it is evident that, the model has been calibrated reasonably well. Among the state variables presented here, prediction of CBOD results found to be better with relative error of 8%, considering the fact that 20 % deviation is normally acceptable in standard river water quality modeling practices.

References

http://www.dailytimes.com.pk/default.asp?page=story_-2002_pg3_8

Brown, L.C., and Barnwell, T.O., Jr. 1987. "The Enhanced Stream water quality Models, Qual-2E and Qual-2E UNCAS: *Documentation and Users Manual.*" EPA /600/3-87/007, Envir. Res. Lab., Envir. Protection Agency (EPA), Athens, Ga.

Children's Misconceptions about Science. Operation Physics, *American Institute of Physics* (September 1998).

Cristi Cave. How a River Flows. Stream Biology and Ecology.

Chitale, M.A.1992.Population and Water Resources of India (In) *The Inevitable Billion Plus: Science, Population and Development, An Exploration of Inter connectivities and Action Possibilities in India, Umesh Communications.* Pune. pp 111-117. Edited by Vasant Gowariker.

Central Pollution Control Board, 2000. Report on Tank Irrigation in India Ministry of Environment and Forests, New Delhi.

Central Pollution Control Board, 1999. A Report on MINARS GOI New Delhi.

Central Water Commission, 1997. Availability of *Water in India.*

Fermat's last theorem, Simon Singh, 1997

Matt Rosenberg (2006-06-08). Do All Rivers Flow South?

Matt Rosenberg. Rivers Flowing North: Rivers Only Flow Downhill; Rivers Do Not Prefer to Flow South.

Mc Cutcheon., S.C. 1989. Water Quality Modelling. Vol.-I. *Transport and Surface Exchanges in Rivers.* CRC press. Inc., 103-110.

Nezette Rydell 1997. Re: What determines the direction of river flow? Elevation, Topography, Gravity??. *Earth Sciences.*

Tata Energy Research Institute (TERI), 1998. *TERI Energy Database and Directory Yearbook (TEDDY), 1996-96.* New Delhi.

Tata Energy Research Institute (TERI), 2001. *TERI Energy Database and Directory Yearbook (TEDDY), 1996-96.* New Delhi.

Thomann., R.V., and Muller., J.A. 1982 " Verification of Water Quality Models." *Journal of Environmental Engineering,* Vol.108, No.EE 5, 923-940.

Thomann, R.V. and J.A. Mueller. 1987. Principles of Surface Water Quality Modeling and Control. New York: Harper & Row.

William, B.M., George, L. B., Thomas, M.G., and Kay, M. 1992. *Hand Book, "Stream Sampling for Waste Load Allocation Application."* US EPA, Office of Research and Development, Washington, D.C 20460.

William C. Philips 1991. Earth Science Misconceptions.

❑❑❑

Chapter 7

Modeling: Coastal and Estuaries

7.1 Introduction

The coast is one of the most important boundaries in the world. It separates the marine part of the world from the continents. It separates the salt-water communities from fresh water communities; it separates land life from water borne life but amphibians living both environments (Fig. 7.1).

Some of the marine organisms move towards estuary for reproduction with less water salinity. The coastline as such is not a stable. It fluctuates in place over considerable distances due to erosion and sedimentation, but also due to the relative change of sea level with respect to land. The coastal zone is thus a very mobile part of the earth from a geographical point of view. Due to its specific geophysical properties, the coast is also a very important area for the development of flora and fauna. There is hardly any other part of the world that houses such a variety of species, both in flora and fauna.

The abundance of opportunities along the coast has also attracted a lot of human activities. The coastal zone has become an indispensable resource for mankind. It provides excellent opportunities for fisheries, housing, recreation, transport, defense purpose, industrial activities and many others. The influx of man with their social and economic activities, their needs for safe housing, food, drinking water, and with the tremendous quantities of waste they are producing poses at the same time a serious threat to the unique features of the coastal zone. It is therefore necessary to study this valuable region so as to prevent that competing activities are gradually causing depletion of the resources in the coastline. The following figures (Fig. 7.1and 7.2) show the typical coastal area of rocky coast and sandy coast.

Fig. 7.1: Rocky Shore

Fig. 7.2 Sandy Shore

Coastal geography is the study of the dynamic interface between the ocean and the land, incorporating both the physical geography (*i.e* coastal geomorphology, geology and oceanography) and the human geography (sociology and history) of the coast. It involves an understanding of coastal weathering processes, particularly wave action, sediment movement and weather, and also the ways in which humans interact with the coast.

7.2 Coastal Processes

The waves of different strengths that constantly hit against the shoreline are the primary movers and shapers of the coastline. Despite the simplicity of this process, the differences between waves and the rocks they hit result in hugely varying shapes. The effect that waves have depends on their strength. Strong, also called destructive waves occur on high energy beaches and are typical of Winter. They reduce the quantity of sediment present on the beach by carrying it out to bars under the sea. Constructive, weak waves are typical of low energy beaches and occur most during summer. They do the opposite to destructive waves and increase the size of the beach by piling sediment up onto the berm.

One of the most important transport mechanisms results from wave refraction. Since waves rarely break onto a shore at right angles, the upward movement of water onto the beach occurs at an oblique angle. However, the return of water is at right angles to the beach, resulting in the net movement of beach material laterally. This movement is known as beach drift. The endless cycle of swash and backwash and resulting beach drift can be observed on all beaches.The wide expanse of sand and low waves, even though a storm is blowing onshore, indicate that this coastline at Rhossili in Wales is a low energy shoreline.

Probably the most important effect is longshore drift (LSD), the process by which sediment is continuously moved along beaches by wave action. LSD occurs because waves hit the shore at an angle, pick up sediment on the shore and carry it down the beach at an angle. Due to gravity, the water then falls back perpendicular to the beach, dropping its sediment as it loses energy. The sediment is then picked up by the next wave and pushed slightly further down the beach, resulting in a continual movement of sediment in one direction.

This is the reason why long strips of coast are covered in sediment, not just the areas around river mouths, which are the main sources of beach sediment. LSD is reliant on a constant supply of sediment from rivers and if sediment supply is stopped or sediment falls into a submarine canals at any point along a beach, this can lead to bare beaches further along the shore.

LSD helps create many landforms including barriers, bay beaches and spits. In general LSD action serves to straighten the coast because the creation of barriers cuts off bays from the sea while sediment usually builds up in bays because the waves there are weaker, while sediment is carried away from the exposed headlands. The lack of sediment on headlands removes the protection from waves them and makes them more vulnerable to weathering while the gathering of sediment in bays protects the bays from further erosion and makes them pleasant recreational beaches.

7.3 Atmospheric Processes

- Wind coming up the beach, picks up sand and moves it up the beach to form sand dunes.
- Rain hits the shore and erodes rocks and carries weathered material to the shoreline to form beaches.
- Warm weather can make for easier for chemical and biological processes to occur. In tropical areas some plants and animals protect stones from weathering, while others actually eat away at the rocks.
- Heat that is sometimes below the freezing point and sometimes above it results in freeze thaw weathering, while weather more than a few degrees below freezing point creates sea ice.

7.4 Biological Processes

Some of the organisims which are live in coastal ecosystems are shown in figures (Fig. 7.3) In tropical regions in particular, plants and animals not only affect the weathering of rocks but are a source of sediment themselves. The shells and skeletons of many organisms are of calcium carbonate and when this is broken down it forms sediment, limestone and clay.

Fig. 7.3 Coastal living organisms

7.5 Chemical Processes

The main chemical process on beaches is salt-crystal growth. Wind carries salt spray onto rocks, where it is absorbed into small pores and cracks within the rocks. There the water evaporates and the salt crystallises, creating pressure and often breaking down the rock. In some beaches calcium carbonate is able to bind together other sediments to form beachrock and in warmer areas dunerock.

7.6 Sea Level Changes

Because the sea level on earth regularly rises and falls due to climatic changes. During cold periods more of the Earth's water is stored as ice in glaciers while during warm periods it is released and sea levels rise to cover more land. Sea levels are currently quite high, while just 18,000 years ago during the Pleistocene ice age they were quite low. Global warming may result in further rises in the future, which presents a risk to coastal cities as most would be flooded by only small rises. As sea levels rise fjords and rias form. Fjords are flooded glacial valleys and rias are flooded river valleys. Fjords typically have steep rocky sides, while rias have dendritic drainage patterns typical of drainage zones. As tectonic plates move about the Earth they can rise and fall due to changing pressures and the presence of glaciers. If a beach is moving upwards relative to other plates this is known as isostatic change and raised beaches can be formed.

7.7 Coastal Landforms

If the coast suddenly changes direction, especially around an estuary, spits are likely to form. LSD pushes sediment along the beach but when it reaches a turn the LSD does not always easily turn with it, especially near an estuary where the outward flow from a river may push sediment away from the coast. The area may be also be shielded from wave action, preventing much LSD. On the side of the headland receiving weaker waves, shingle and other large sediments will build up under the water where waves are not strong enough to move them along. This provides a good place for smaller sediments to build up to sea level. The sediment, after passing the headland will accumulate on the other side and not continue down the beach, sheltered both by the headland and the shingle.

Slowly over time sediment simply builds on this area, extending the spit outwards, forming a barrier of sand. For example, Kakinada bay, a sand spit is formed due to sediment transport is northword in the order of 0.4 to 0.5 million cubic meter per year. Another example at Kalpakkam, the sand spit is formed due to LSD. Once in a while, the wind direction will change and come from the other direction. During this period the sediment will be pushed along in the other direction. The spit will start to grow backwards, forming a 'hook'. After this time the spit will grow again in the original direction. Eventually the spit will not be able to grow any further because it is no longer sufficiently sheltered from erosion by waves, or because the estuary current prevents sediment resting. Usually in the salty but calm waters behind the spit there will form a salt marshland. Spits often form around the breakwater of artificial harbours requiring dredging.

Occasionally, if there is no estuary then it is possible for the spit to grow across to the other side of the bay and form what is called a bar, or barrier. Barriers come in several varieties, but all form in a manner similar to spits. They usually enclose a bay to form a lagoon. They can join two headlands or join a headland to the mainland. When an island is joined to the mainland with a bar or barrier it is known as a tombolo. This usually occurs due to wave refraction, but can also be caused by isostatic change, a change in the level of the land (Chesil Beach). An example of this is along the Holderness coastline.

7.8 Coastal Water Pollution

The various sources that cause pollution of coastal water arise from the discharges of domestic sewage and industrial effluents, discharge of domestic sewage and rubbish from ships, transportation of oil and exploration and exploitation of the coastal bed and ocean floor. Since the tendency of self purification of the sea is high due to different factors such as dilution, dispersion in temperature etc., and pollution of the coastal water is not considered as serious as river pollution.

India is one of the growing industrial countries in the world. National inventory (Report 1995) of industries revealed that there are 8432 large and medium industries existing in the country but the number was about half a decade ago. With regard to industrial settlement in the country the leading states are Goa, Maharashtra, Karnataka, Tamilnadu, Andhra Pradesh, Uttar Pradesh, Madhya Pradesh, Punjab, Rajasthan and West Bengal. Where Tamil Nadu the leader in the east coast (20%) and Maharashtra in the west (16%). With regard to distribution of large and medium industries in the littoral states, there are, in total, 491 industries located in both the coast of which 52% in the west coast and the remaining 48% in the east. A study was made in 1994/95 to assess the pollution potential of coastal industries in the country. Under this study, a coastal landmass of 25 km in shore was considered. It has been observed that there is a good concentration of large and medium industries in the narrow coastal belt. Statewide coastal length, area and number of large and medium industries as also aquacultural farms located in the coastal area with in 25 km land ward along with the number of industries under 17 categories are given in table 5.1. It may be observed that 308 large and medium industries are located within the coastal belt of the country which are likely to discharge their liquid and solid wastes directly or indirectly into the coastal waters. Out of these, 233 industries are located in the west coast and 75 on the east coast. In addition, from the available information, there are 138 aquaculture farms predominantly for shrimp farming located along the east coast alone which is more than 75% of the total number of such farms in the country.

Table 7.1 Details of Industries in the Coastal India

S. No.	Coastal State	Coastal length (km)	Area up to 25km from shoreline	Industries	Aquaculture farms
1.	Gujarat	1663	41,575	35(29)	-
2.	Maharashtra	720	18,000	167(38)	-
3.	Goa	140	3,500	2(1)	
4.	Karnataka	290	7,250	3(2)	-
5.	Kerala	560	14,000	26(12)	-
6.	TamilNadu	860	21,500	30(15)	20
7.	Pondicherry	24	600	4 (4)	-
8.	Andhra Pradesh	930	23,250	30(24)	88
9.	Orissa	450	11,250	4(2)	20
10.	West Bengal	200	5,000	7(4)	10

Western Coast

Gujarat

Of all maritime state, the state of Gujarat has the largest coastline. Important rivers outfall are Sabarmati, Mahi, Narmada, Tapi, Purna, Saraswati, Indola, Ambika, Par, Kolak and Daman Ganga.

Not less than 35 important industries including seven industrial estates are located ar or near the coast of which chemicals;

petrochemicals, refinery, fertilizers, pharmaceuticals, pesticides textiles and power plants are predominating. Major industrial locations are Atul, Utran, Hazira, Vapi and Vadodara. Besides release of effluents from coastal industries and untreated domestic wastes into the coastal water, the port activities at Kandla, Porbandar,Okha and Surat including dredging, navigation and transport of crude oil are the sources of potential threats to the coastal waters.

Maharashtra

The coastal zone of the state is traditionally known as KONKAN. The coastal zone is mainly rocky and broacken by small creeks and fringed with islands. Mentionable small rivers outfall are Ulhas, Patalganga, Vogeswar, Amba, Kundalika, Savitri, Vasishtha. Major industial locations are Mumbai- Thane complex and Ratnagiri. Out of a total 167 industries major types of industries like chemical, petrochemicals, pharmaceuticals, textiles, engineering and metallurgy are located in these regions. Besides industrial activities, the port related activities at Mumbai,JNPT and Nhavasheva and transport of crude oil also contribute substantially to the total contaminant load in the coastal waters.

Goa

In general, the Goa coast is sandy. The prominent river outfalls are Mandovi and Zuari. The prime importance of the coast is that it has an extensive beach resource of international tourist interest. Only one pharmaceutical unit and one sewage treatment plant are located at Panaji and Corlim. One of the major concern in the system generating from the mine rejects dumped along the shore which are ultimately transported into coastal seas.

Karnataka

The coastline is well defined and almost straight but broken at numerous points by rivers, creeks, and bays. River outfalls in the coast are formed by small rivers namel, Kalinadi, Bedti, Sharavati, Swarnanadi. The coast is almost free from industries. Only three industries of chemicals, fertilizers and Iron ore are located at Mangalore and Karwar region. Besides waste generated from 12 coastal towns. The port related activities of New Mangalore and Bharkal including a number of fishing harbours also form the potential sources of coastal pollution.

Kerala

The coastal planes of kerala is endowed with brackish water lagoons interconnected with natural and man-made canals as also 41 rivers outfalls. The conspicuous feature is that it is a small state but has utilizable coastline of about 14km per 100 sq.km land area. Major industries in the coastal zone are in the category of chemical, petrochemical and fish processing mainly located in the three regions like Calicut, Cochin, Udhyogmandal-Ambala mugal area. Besides industrial waste and municipal waste generated from 14 coastal towns, heavy traffic of marine vessels for fishing, port activities at Cochin, Alleppy, Calicut and Defense related activities at Cochin also form the variable sources of pollution in the coastal water.

Eastern Coast

Tamil Nadu

The coastline of Tamilnadu is mostly facing the Bay of Bengal to the east with only a smaller part facing the Arabian Sea. The plain is mostly alluvial formed by the important tributaries namely Vellor and Coleroon has formed an extensive deltaic mass at the coast fringed with mangroves. Other important river outfalls are Palar, Cheyyar, Amarabati, Pennar and Tamraparni. The coastal belt is predominantly occupied by the aquaculture farms and sea food processing industries. Other industries like fertilizers, petro and agro chemicals and chemicals are mainly located at Chennai, Ennore, Cuddalor, Avadayaikoil, Ottapidaram and Tuticorin. Besides industries and municipal waste, port related operation with busy traffic of marine vessels at Chennai, Tuticorin, Karaikal including oil transport as allows the waste of aquaculture farms are increasingly posing threats to the coastal water quality.

Puducherry

The union territory of Puducherry is surrounded by Tamilnadu state on the landward side and Bay of Bengal in the east. The coastline is very small. Only four large and medium industries are located along the coast of which a distillery and paper mill contribute substantially to the total contaminant load in the coastal water.

Andhra Pradesh

The coastal geomorphology Andhra Pradesh coast is characterized by sandrifts in the north, rocky cliff in the south with way headlands, the prominents of which is known as 'Dolphin'noise' behind this occure coastal march with mud flats and salt pans. Two broad delta have been formed by the Godavari and Krishna rivers. Other important rivers are Pennar, Nagavalli and Vamsadhara. Major types of coastal industries namely sugar, distilleries, chemicals, refinery, fertilizers, paper and large number of aquacultureal forms and marine food processing industries are located in the regions-Bhogapuram, Bhimunupatram, Visakhapatnam, Rambillimondal, Nakkapalli-mondal, Kothapatnam and Allura. In addition, the port activites at Visakhapatnam and Kakinada also form the potential sources of coastal pollution.

Orissa

The coast planes of Orissa has two districts topographical features. The zone from the state of West Bengal to the north of Pola deep is concave and very shallow with sandy beaches. The area is characterized by major river outfalls namely Subarnarekha, Brahmanii, Baitarani and the Mahanadi. The remaining part from near Paradip to the state of Andhra Pradesh is protrudingly exposed to the sea with rocky beds at places. The coastal area is relatively free from industries except a phosphate and distillery industries at Paradip and a chemical and rare-earth near Gopalpur. Besides industrial and municipal wastes from coastal towns, the other important potential sources of coastal pollution are port activities at Gopalpur and Paradip. Intensive mechnised fishing and waste of good number of aquaculture farms.

West Bengal

The coast of West Bengal, unlike the remaining coastal fronts of India, exhibits a distinctive physiographical feature. The Ganga, Brahmaputra river system, transporting huge suspension load, has created a vast inter-tidal deltaic mass 'Sundarbans' the largest delta in the world with extensive mangrove forests. The Hugli-Malta river system and six others esturies with tidal creeks and channels forms and extensive network in the lower Bengal. Although the state is very rich in industrial settlement, the coastal belt in almost devoid of industries. However an industrial complex at Haldia Hoogly about 40km from the coastline has grown up. Like remaining coast in the east, the coastal aquacultural farming has also been a prominent activity along the West Bengal coast. The major sources of pollution in the West Bengal coast are the sewage of Calcutta metropolis and port operations at Haldia and Calcutta including growing traffic of oil tankers.

As regards liquid and solid waste generation, the total quantity of about 1.35 million cubic meters of effluent and about 34,500 tons of solid waste are generated by the coastal industries every day. State wise coastal industries and their respective waste generation are given in table 5.2.2. It may be observed that the largest concentration of effluent in industries is in Maharashtra followed by Gujarat, Tamil Nadu, Andhra Pradesh and Kerala. On the other hand, the largest concentration of aquaculture farms is in Andhra Pradesh, Tamil Nadu and Orissa in the east coast. Out of the total solid waste of 34,476 tons generated per day about 44% in the west coast though the number of industries there is about three tons that in the east coast where 56%. This disproportionate generation of waste may be attributed to the power plants and steel mills located in the states of Tamil Nadu and Andhra Pradesh. In the west coast, generation of large quantities of solid waste in Gujarat is also due to the similar reason.

In the generation of liquid waste, out of the total quantity of 1.35 million cubic meters per day, about 0.85 million cubic meters per day is from west coast industries, in which by far the largest quantity is generated in Gujarat. The east coast industries generated the total quantity of about 0.5 million cubic meters per day, in which Andhra Pradesh alone contribute 0.35 million cubic meters per day, followed by Tamil

Table 7.2. State wise Coastal Industries Waste Generation

State	Types of pollutant	Industries (Nos)	Industrial Effluent (million m^3/day	Industrial Solid waste (T/day)	Aquaculture farms(Nos)	Aquaculture farms waste (million m3/day
Gujarat		35	0.566	95066	-	-
Maharashtra		167	0.080	2628	-	-
Goa		2	0.012	1.4	-	-
Karnataka		3	0.043	76	-	-
Kerala		26	0.515	2431	-	-
TamilNadu		30	0.125	9112	20	0.253
Pondicherry		4	0.006	1.25	-	-
Andhra Pradesh		30	0.350	7191	88	2.12
Orissa		4	0.001	3505	20	-
West Bengal		7	0.022	25	10	-
East coast		75	0.50	19834	138	2.37
West coast		233	0.85	14642	-	-

Nadu that ie 0.125 million cubic meters per day. The liquid waste generation from aquacultural farms from east coast only, in absent of data on west coast , is the tune of 2.37 million cubic meters per day, out of which Andhra Pradesh has the lion share of about 2.12 million cubic meters per day.

From the pollution point of view, the 17 categories of industries (of highly polluting in nature) comprise a total of 131 in the coastal area. Out of which 38 are located in Maharashtra, 29 in Kerala and remaining 13 being distributed in other 4 states and one Union Territory. The state wise distribution of 17 categories are given in table 7.2 (Manivanan, 2006).

Water quality models can simulate plume movement and dilution from single or multiple discharge points as well as background concentration fields, initial dilution, near and far field dilution/ concentration, DO/BOD/nitrification, pH, degradation, coliforms, suspended sediment, adsorption and residence time.

- BEL1D/2D - Estuary modelling
- BELPLUME - 3D Water quality prediction, estuaries and coastal areas
- FLUENT - 3D River, canal and estuary modelling
- MIKE 21 - 2D Hydrodynamic and water quality modeling.
- CORMIX - 3D Hydrodynamic and water quality modeling in coastal area
- TELEMAC - Flexible Mesh hydrodynamic and water quality modeling in coastal area

Longer timescale modelling to simulate the fate of chemicals is also undertaken, including adsorption/desorption, flushing, sediment movement, degradation, volatilisation and persistence. Effect modelling techniques are currently under development at BEL, to predict acute and chronic toxic zones and the effects on aquatic populations.

Model validation and sensitivity/uncertainty tests form an important part of every water quality modelling study. This ensures that the model is applicable over the full range of environmental conditions that need to be considered.

Dispersion measurement - dye release and dilution monitoring

- Salinity, temperature, pH and dissolved oxygen measurement
- Water quality measurements (e.g. ammonia, nitrate, metals)
- Sediment quality measurements (e.g. metals, organics)

7.9 Estuary Modeling

The model set up for the modeling studies of the Kokemäenjoki estuary has been completed and first model runs have been carried out. The model is forced by DIN, DIP as well as phytoplankton loading from the Kokemäenjoki on a monthly basis for DIN/DIP and for the summer months (June-August) for phytoplankton for the years 1995-1999. The results of the simulation experiments are compared with observations which are derived from the costal area off Pori. In agreement with observations, results show a pronounced spring bloom which occurs end of April/begin of March followed by moderate phytoplankton concentrations during summer and autumn.

The concentration of DIN and DIP decrease in late spring due to the uptake of phytoplankton and increase again in autumn and winter due to demineralization and mixing. The model lacks in simulating the high phytoplankton concentrations in spring as well as the low concentrations of nutrients in summer which can be attributed to (i) missing data for phytoplankton loading in spring, (ii) too coarse time resolution for loading and (iii) too coarse horizontal and vertical model resolution (5 km). Besides the evaluation of possible model improvements, in future, reasonable scenarios on nutrient loading will be carried out. The effect of nutrient reduction on phytoplankton growth and seasonal variations of phytoplankton and nutrients will be investigated.

7.10 A Case Study on East Coastal Area

Kalpakkam is located on the east coast of India, about 75 km south of Chennai in Tamil Nadu State. The mathematical model for thermal and littoral drift studies is carried out in which seasonal/annual nearshore wave climate was determined, seasonal/annual littoral drift rates and their cross-shore distributions were estimated, shoreline changes due to proposed development were computed.

7.10.1 Longshore Drift Rates Computed

Monthly littoral drift along the Chennai coast was computed by Prasad and Reddy (1988). Deep water wave data of 10 years period (1970-79) compiled by India Meteorological Department was used and breaker wave heights (H_b) and breaker wave angles (a_b) along the Chennai coast were computed with the help of wave-refraction diagrams.

The littoral drift is directed towards north during nine months of the year (February to October), while the drift is towards the south during three months (November, December and January). The monthly littoral drifts are minimal for February and October(around 400,000 m^3 per month) while a maximum of about 420,000 m^3 per month occurs during November and in December it was estimated to be 360,000 m^3 per month. The total northerly and southerly drifts in a year were estimated to be 1,550, 000 m^3 and 1,010,000 m^3 respectively. Thus the annual gross and net(towards north) drifts become 2,560,000 m^3 and 540,000 m^3.

The total amounts of material deposited to the south of the port entrance from February to September, evaluated for each year from 1973–1978 indicated the amounts of accretion are highly variable from year to year. A maximum deposition of about 1,100,000 m^3 occurred during 1976, while a minimum deposition of about 400,000 m^3 occurred during 1974. On average about 700,000 m^3 of material is deposited in the region during February to September. It is indicated that as the computed total northerly drift is 1,550,000 m^3, it is evident that only part of the material is deposited in the region, while the rest is either bypassed to the northern shore or carried offshore by the possible rip currents in the region.

Using deep water wave data, Chandramohan et al(1990) estimated northward drift of 1.03x 10^6 m^3 and southward drift of 0.68x10^6 m^3 with net drift(towards north) to be 0.343x 10^6 m^3 along Madras coast. They indicate drift towards north during March to October and towards south during November to February.

These researchers reported that the drift is Southward during November- February. Maximum northerly drift occurs during June and July whereas southerly drift is maximum in January and February. About 0.08 to 0.14 million m^3 of sand is drifted towards south in a year. Along the coastal stretches of Grid I(Lat. 9-12 N, Long. 80-83 E) and Grid II(12-15 N, 81-84 E)a net transport of 0.35-0.4 million m^3 of sand towards north occurs. (Natesan and Subramanian, 1994)

Shoreline stability studies are carrird out by NIO, Goa. They computed longshore transport along Kalpakkam coast using wave parameters and longshore currents. On an average, for the whole

Kalpakkam coast, the annual gross northerly transport is about 1.44 million m^3 and southerly transport is around 0.35 million m^3. The average net transport towards north is about 1.09 m^3.

On the basis of wave data obtained by visual observations by Madras port authorities seasonal transport rates computed are reported by Sunder(2002) for the years 1974-79 from which the average northward and southward transport is 2.44 million m^3 and 1.37 million m^3 with the net northward transport to be 1.06 million m^3.

LITPACK model is used to estimate seasonal/annual littoral drift rates and its distribution on the profile normal to shoreline. The model assumes the depth contours parallel to the coast. The shoreline at Kalpakkam is oriented approximately in SSW-NNE direction. Profiles are chosen for computation of littoral drift which are at Kalpakkam each profile is discretised into 150 grid points with grid size of 10 m.

Besides the inshore wave climate and bathymetry along a cross-shore profile, grain size distribution over the profile is also required. At the shore, grain size is observed to be of the order of 0.3 mm. Hence d_{50} over the profiles is assumed to vary from 0.3 to 0.1 mm. The bed roughness is adjusted to get the annual northward and southward transports of the order of 1.0 and 0.5 million m^3 . On the basis of seasonal wave climate the model computes seasonal transport rates.

A total reach of 10 km of the shoreline is selected for the model with the consideration that the effect of the impact of the outflow may propagate for about 5 km of the shoreline on the either side of the outlet. With this shoreline the shoreline evolution model is run for 3 years period. As the net transport is towards north, deposition on south side and erosion on the north side of the barrier is observed. After a period of 3 years, the maximum cross-shore advancement of the shoreline on the south side would be of the order of 30 m and effect of the advancement would be felt for about 2-3 km. Similarly after 3 years, maximum cross-shore recession of the shore line on the north side would be of the order of 20 m with its longshore effect for about 2-3 km

From the above discussions it is seen that there is considerable variation in estimation of littoral drift in the region due to the uncertainties involved. However, in general, the annual rate of northward and southward transport at the Chennai coastline are of the order of 1.0 and 0.5 million m^3 respectively. Hence the northward and southward transports for Kalpakkam coastline are assumed to be

1.0 and 0.5 million m^3 so that net and gross transports become 0.5 and 1.5 million m^3.

References

Chandramohan P., Nayak. B.U. and Raju V.S., 1990 'Longshore sediment model for south Indian and Shrilankan coasts' *ASCE Jl.of Waterway, Port, Coastal and Ocean Engineering*, Vol.116, pp.408-424.

Manivanan, 2006 'Recycling of industrial effluents' NIPA, New Delhi

Natesan U and Subramanian S.P., 1994 ' Study of transport processes along Tamil nadu coast', *Proc. Indian National Conference on Harbour and Ocean Engineering* (INCHOE-94), Vol.1 pp. D-103-D-108.

Prasad K.V.S.R and Reddy B.S.R 1988 ' Near-shore sediment dynamics around Madras port, India', *ASCE Jl. Of waterway, port, coastal and Ocean Engineering* , Vol.114. No.2, pp.206-209.

Sunder .V., 2002 ' Littoral drift in Harbour and Coastal Engineering (Indian Scenario)', Vol II. Ed. S. Narasimhan and S. Kathiroli, NIOT, Chennai.

❑❑❑

Chapter 8

Modeling: Lakes and Reservoirs

8.1 Introduction

A lake is a body of water or other liquid of considerable size surrounded entirely by land (Fig. 8.1). A vast majority of lakes on Earth are fresh water, and most lie in the Northern Hemisphere at higher latitudes. In ecology, the environment of a lake is referred to as lacustrine.

Fig 8.1 A Typical Lakes

Large lakes are occasionally referred to as "inland seas" and small seas are occasionally referred to as lakes. Smaller lakes tend to put the word "lake" after the name, as in Green Lake, while larger lakes often invert the word order, as in Lake Ontario, at least in North America.

The term lake is also used to describe a feature such as Lake Dal, which is a wet basin most of the time but may become filled under seasonal conditions of heavy rainfall and snowfall. Many lakes are artificial and are constructed for hydro electric power supply, recreational purposes, industrial use, agricultural use, or domestic water supply.

The natural results of erosion will tend to wear away one of the basin sides containing the lake, such as the shores of Lake Baikal in Russia which is estimated to be 25-30 million years old. There are a number of natural processes that can form lakes. A recent tectonic uplift of a mountain range can create bowl-shaped depressions that accumulate water and form lakes. The advance and retreat of glaciers can scrape depressions in the surface where lakes accumulate; such lakes are common in Scandinavia, Patagonia, Siberia and Canada. Lakes can also form by means of landslides or by glacial blockages. An example of the latter occurred during the last ice age in the US state of Washington, when a huge lake formed behind a glacial flow; when the ice retreated, the result was an immense flood that created the Dry Falls at Sun Lakes, Washington.

Salt lakes can form where there is no natural outlet or where the water evaporates rapidly, and the drainage surface of the water table has a higher than normal salt content. Examples of salt lakes include Great Salt Lake, the Caspian Sea, the Aral Sea and the Dead Sea.

Small, crescent-shaped lakes called oxbow lakes can form in river valleys as the result of meandering. The slow moving river forms a sinuous shape as the outer side of bends are eroded away more rapidly than the inner side. Eventually a horseshoe bend is formed and the river cuts through the narrow neck. This new passage then forms the main passage for the river and the ends of the bend become silted up, thus forming a bow shaped lake.

Lake Vostok is an subglacial lake in Antarctica, possibly the largest in the world. The pressure from ice and the internal chemical composition means that if the lake are drilled into, it may result in a fissure which would spray in a similar fashion to a geyser.

Some lakes, such as Lake Baikal and Lake Tanganyika lie along continental drift zones, and are created by the crust's subsidence as two plates are pulled apart. These lakes are the oldest and deepest in the world, and may be destined over millions of years to become oceans. The Red Sea is thought to have originated as a rift valley lake.

8.2 Types of Lakes

The following lakes are classified based on their nature and geographical location.

- *Periglacial:* Part of the lake's margin is formed by an ice sheet, ice cap or glacier, the ice having obstructed the natural drainage of the land.
- *Subglacial:* A lake which is permanently covered by ice. They can occur under glaciers and ice caps or ice sheets. There are many such lakes, but Lake Vostok in Antarctica is by far the largest. They are kept liquid because the overlying ice acts as a thermal insulator retaining energy introduced to its underside by friction, water percolating through crevasses, by the pressure from the mass of the ice sheet above or by geothermal heating below.
- *Artificial,* also called a reservoir: A lake created by flooding land behind a dam, by human excavation, or by the flooding of an open pit mine. Some of the world's largest lakes are reservoirs. Husain Sagar is a reservoir in India built in 1562.
- *Endorheic,* also called *terminal* or *closed*: A lake which has no significant outflow, either through rivers, or underground diffusion. Any water within an endorheic basin leaves the system only through evaporation. These lakes are most common in desert locations, such as Lake Eyre in central Australia or the Aral Sea in central Asia.
- *Meromictic:* A lake which has layers of water which do not intermix. The deepest layer of water in such a lake does not contain any dissolved oxygen. The layers of sediment at the bottom of a meromictic lake remain relatively undisturbed because there are no living organisms to stir them up.
- *Oxbow:* A lake which is formed when a wide meander from a stream or a river is cut off to form a lake. They are called oxbow lakes due to the distinctive curved shape that results from this process.
- *Rift lakes:* A lake which forms as a result of subsidence along a geological fault in the Earth's tectonic plates. Some examples are the Rift Valley lakes of eastern Africa.
- *Underground:* A lake which is formed under the surface of the Earth's crust. Such a lake may be associated with caves and aquifers and springs.

- *Crater:* A lake which forms in volcanic calderas or craters after the volcano has been inactive for some time. Water in these types of lakes may be fresh, or highly acidic, and may contain various dissolved minerals. Some also have geothermal activity, especially if the volcano is merely dormant rather than extinct.
- *Former:* A lake which is no longer in existence. Such lakes include prehistoric lakes, and lakes which have permanently dried up through evaporation or human intervention. Owens Lake in California, USA is an example of a former lake. Former lakes are a common feature of the Basin and Range area of south-western North America.
- *Shrunken:* Closely related to former lakes, a shrunken lake is one which has drastically decreased in size over geological time. Lake Agassiz is a good example of a shrunken lake, which covered much of central North America. Some notable remnants of this lake are Lake Winnipeg, and Lake Winnipegosis.

8.3 Limnology

Limnology is the study of inland bodies of water and related ecosystems, and divides lakes in three zones: littoral zone, which is a sloped area that is close to land; photic or open-water zone, where sunlight is abundant; and deep-water profundal or benthic zone, where little sunlight can reach. The depth which light can reach in lakes depends on the density and motion of particles. These particles can be sedimentary or biological in origin and are responsible for the color of the water. Decaying plant matter, for instance, may be responsible for a yellow or brown color, while algae may result in greenish water. In very shallow water bodies, iron oxides make water reddish brown.

Biological particles are algae and detritus. A sediment particle is in suspension if its weight is less than the random turbidity forces acting upon it. The turbidity is a decisive factor in the transparency of the water. Bottom dwelling detritivorous fish are responsible for turbid waters, because they stir the mud in search for food. Piscivorous fish eat plant eating fish, thus increasing the amount of algae. The light depth or transparency is measured by using a Secchi disk. This is a 20 cm disk with alternating white and black quadrants. The depth at which the disk is no longer visible, is the Secchi depth, and is a measure for transparency. It is commonly used to test eutrophication.

A lake moderates the surrounding region's temperature and climate because water has a very high specific heat capacity (4,186 $J \cdot kg^{-1} \cdot K^{-1}$). In the daytime, the lake can cool the land beside it with local winds, resulting in a sea breeze; in the night, it can warm it, forming a land breeze.

A lake may be infilled with deposited sediment, and gradually, the lake becomes a wetland, such as a swamp or marsh. An important difference exists between lowland and highland lakes: lowland lakes are more placid, are less rocky/more sedimentary, have a less sloping bottom, and generally contain more plant life. Large water plants accelerate this closing process significantly because they trap sediment. Turbid lakes, and lakes with much plant eating fish, tend to disappear slower. A "disappearing" lake typically has a water's edge with extensive plant mats.

They become a new habitat for other plants and animals, many of which are very rare. Gradually, the lake closes, and young peat may form, forming a fen. In lowland river valleys , the presence of peat is explained by the closing of historical oxbow lakes. In the very last stages of succession, more trees would grow in, eventually turning the wetland into a forest. Some lakes can also disappear seasonally; they are called intermittent lakes and are typical of karstic terrain. A prime example of this is on 3rd June 2005 in Bolotnikovo, Russia, a lake called White Lake vanished in a short period of time.

8.4 Notable Lakes

- The *largest* lake in the world by surface area is the Caspian Sea. With a surface area of 394,299 km^2, it has a surface area greater than the next six largest lakes combined.
- The *deepest* lake is Lake Baikal in Siberia, with a bottom at 1,637 m (5,371 ft.) and is the world's largest freshwater lake by volume.
- The world's *oldest* lake is Lake Baikal, followed by Lake Tanganyika (Tanzania).
- The world's *highest* lake is the Lhagba Pool in Tibet at 6,368 m.[2]
- The world's *highest* commercially navigable lake is Lake Titicaca in Bolivia at 3,812 m. It is also the largest freshwater lake in South America.

- The world's lowest lake is the Dead Sea at 418 m (1,371 ft) below sea level. It is also one of the lakes with highest salt concentration.
- The largest freshwater lake by surface area, and third largest by volume, is Lake Superior with a surface area of 82,414 km^2. However, Lake Huron and Lake Michigan form a single hydrological system with surface area 117,350 km^2, sometimes designated Lake Michigan Huron. All these are part of the Great Lakes of North America.
- The largest freshwater-lake island is Manitoulin Island on Lake Huron, with a surface area of 2,766 km^2. Lake Manitou, located on Manitoulin Island, is the largest lake on a freshwater lake island.
- The largest lake located on an island is Nettilling Lake on Baffin Island.
- The largest lake in the world that drains naturally in two directions is Wollaston Lake.
- Lake Toba on the island of Sumatra is located in what is probably the largest resurgent caldera on Earth.
- The largest lake located completely within the boundaries of a single city is Lake Wanapitei in the city of Greater Sudbury, Ontario, Canada. Before the current city boundaries came into effect in 2001, this status is held by Lake Ramsey, also in Sudbury.
- Lake Enriquillo in Haiti is the only saltwater lake in the world inhabited by crocodiles.
- Chilka Lake is the largest saltwater lake in Orissa, India.

8.5 Objectives & Approach Water Quality Modeling of a Lake

The objectives of this chapter to develop a model of a Lake that could:

1. Simulate the circulation, temperature, and water quality in the lake,
2. Aid in developing a better understanding of lake circulation and water quality and the processes affecting them, and
3. Predict the changes in circulation, temperature, and quality

To meet these objectives, the model is constructed, calibrated, and tested for circulation, water temperature, and water quality for a Lake. The model is constructed to simulate conditions that occurred in the years 2000 through 2003. Model calibration focussed on the years 2000 and 2001; the calibrated model is then tested with data from 2002 and 2003. Hydrologic conditions during this period are variable, from near-normal conditions in most years to extreme drought conditions in winter 2000-2001, which caused the lake to only reach approximately 50% of storage capacity in 2001.

After the model is constructed and calibrated, it was used to simulate scenarios to help predict the water quality changes that might occur in the lake and downstream.

8.6 Analysis and Results

The a Lake model is constructed using CE-QUAL-W2 version 3.12 (Cole and Wells, 2002). CE-QUAL-W2 is a laterally averaged, two-dimensional flow and water-quality model developed by the U.S. Army Corps of Engineers and Portland State University. The simulated dimensions are longitudinal and vertical. The model simulates flow, horizontal and vertical velocities, water temperature, and a suite of water quality constituents. It has been applied to more than 400 lakes and reservoirs around the world, and is well documented and supported. USGS personnel modified the model for this application, adding the capability to simulate a community of zooplankton (Rounds et al, 1999; Rounds and Wood, 2001).

Input data for the a Lake model included lake bathymetry, meteorologic conditions, tributary inflows, tributary temperature and water quality, and lake outflows. Calibrated output included lake hydrodynamics, water temperature, orthophosphate, total phosphorus, ammonia, algae, chlorophyll-*a*, zooplankton, and dissolved oxygen. Other simulated constituents included nitrate, dissolved and particulate organic matter, dissolved solids, and suspended sediment. Two algal groups (blue-green algae, and all other algae) are included in the model to simulate the lake's algal communities. Measured lake stage data are used to calibrate the lake's water balance; calibration of water temperature and water quality relied upon vertical profile data taken in the deepest part of the lake near the dam.

Sensitivity tests are performed to examine the response of the model to specific parameters and coefficients, including the light-extinction coefficient, wind speed, tributary inflows of phosphorus, nitrogen and organic matter, sediment oxygen demand, algal growth rates, and zooplankton feeding preference factors.

The report describes the objectives and results of the modeling work, including a quantification of model performance, a discussion of those processes that influence water quality in the lake, and the results of sensitivity tests. A second report describes the use of the model to predict the effects of various damraise scenarios. In these scenarios, the model is used to assess the changes in temperature and water quality.

8.7 Significant Findings

The following findings are the result of an investigation of a Lake as it existed in 2000-2003. For the results of the model predictions for possible future conditions.

1. Lake levels are highest in late spring and early summer and decreased through the summer and fall as downstream users required water for irrigation, drinking water, flow augmentation, and municipal uses. Lake levels are lowest in November and began to rise with the onset of winter rains. A drought during the winter of 2000-2001 caused the reservoir to not fill in 2001. The annual cycle in lake level is an important factor that affected lake temperature and water quality.

2. Spatial and temporal patterns in water temperature in the Lake are similar in all 4 years modeled in this study. A thermocline developed each year by early summer, isolating cold, dense water near the bottom, below the lake's outlet structure. Withdrawals from the lake, as well as other seasonal factors, tended to draw the thermocline down to the level of the lake's outlet structure by mid-summer. The Lake typically turned over in November and remained uniformly mixed and isothermal until early March, when temperature stratification began. Meteorological factors (solar energy, air temperature, wind) and reservoir operations (lake stage, elevation of the withdrawal) are found to have significant influences on the lake's water temperature.

3. During normal years, dissolved oxygen become depleted in the hypolimnion by late September; during the drought year, this occurred earlier, by late August. Colder temperatures and lake turnover in November reoxygenated the water column and ended hypolimnetic anoxia in each year. Dissolved oxygen levels in the lake are controlled mainly by water temperature and sediment oxygen demand, and to a lesser degree, by algal photosynthesis and respiration.

4. Ammonia concentrations generally are low throughout study period in the the lake. However, in all years studied, accumulation of ammonia in the hypolimnion occurred once dissolved oxygen is depleted. Ammonia concentrations as high as 0.43 mg/L as N are measured in November of 2000.

5. Algae are separated into two groups in the model: blue-green algae, and all other algae. The general algae group had its highest abundance in the spring, due in part to inputs of algae from tributaries and possible resuspension of algal cells during storms. The blue-green algae group tended to bloom in late summer. Orthophosphate concentrations, as well as zooplankton grazing, water temperature, and light, controlled the levels and timing of algal blooms in the model. Concentrations of bioavailable phosphorus appeared to limit the size of the annual blue green algae bloom.

6. While a community of zooplankton is found in the Lake, its interactions with other zooplankton and with the lake's algal communities appears to be significantly more complex than what is represented in the model.

7. The model captured the dominant processes affecting water quality (temperature, dissolved oxygen, nutrients, and algae) in the Lake and simulated the lake's water quality dynamics with sufficient accuracy for the planned purposes of the model. Comparing measured and modeled water temperature profiles for the 4 years simulated, the mean absolute error (MAE) is less than 0.7°C and the Root Mean Square Error (RMSE) is less than 1°C. Comparing measured and modeled dissolved oxygen profiles in the 3 years having reliable data (2000-2002), the model errors typically were less than 1 mg/L.

8.8 Water Quality Modeling in Tropical Reservoirs

An initial issue for consideration related to reservoirs (Fig. 8.2) created as a result of hydropower, water supply and flood control projects is the impacts of such projects on water quality. Some reasons for concern related to water quality include: (1) the increasing use of multipurpose reservoirs; i.e., reservoirs for hydroelectric power, flood control, and/or water supply; (2) the increasing adoption and enforcement of water quality standards in relation to reservoir and downstream water uses; and (3) the documentation of water quality changes based on information collected over the last 25 years (Canter, 1977). Since water quality changes are often rate and temperature dependent, they are of increased concern in tropical reservoirs due to higher temperatures and larger rate constants.

Fig. 8.2 A Typical Reservoirs

A number of key processes within a reservoir can influence water quality. These processes include thermal/density stratification, sedimentation, evaporation, chemical and/or biological cycling, bacterial die-away, and gas or nitrogen supersaturation. Summary points regarding these processes are as follows:

(1) Thermal/density stratification and resultant epilimnetic and hypolimnetic processes (aerobic processes in epilimnion and anaerobic processes in hypolimnion).

   (a) The density of water varies with the temperature of water (Smalley and Novak, 1978).

(b) As a result of density differences, can have density induced flow anomalies in reservoir (for example, warmer water flowing over colder water, colder water flowing under warmer water, or cooler water flowing between warmer and colder water).

(c) Because of the layering effect can get reservoirs which thermally stratify and thus an epilimnion, thermocline, and hypolimnion can form (Smalley and Novak, 1978).

(d) Epilimnion characterized by good oxygen transfer, good mixing and diurnal temperature fluctuations; the thermocline is the transition zone; and the hypolimnion is characterized by low oxygen to anaerobic conditions, limited mixing, and nearly isothermal conditions.

(e) Stratification can lead to both lateral and vertical water quality changes within a reservoir.

(2) Sedimentation of suspended materials which flow into the reservoir; suspended materials inputs are a function of drainage area size, characteristics, rainfall, and other factors:

(a) Sedimentation rate varies as a function of sediment size and other factors such as temperature and turbulence; sedimentation will increase with temperature and decrease with turbulence.

(b) Excessive sedimentation can cause reservoir filling.

(c) Sediments can adsorb dissolved constituents and transport them to the reservoir benthic zone.

(d) Sediments can disrupt benthic organisms and cover previously deposited pollutants.

(3) Evaporation of water from the reservoir will lead to higher concentrations of dissolved constituents; evaporation rates increase with increases in temperature.

(4) Chemical and/or biological cycling of water quality constituents; numerous examples and processes can be cited:

(a) Nitrogen inorganic and organic chemical forms, nitrification/denitrification, and biological uptake.

(b) Phosphorus inorganic & organic chemical forms, adsorption, ion exchange, precipitation, and biological uptake.

(c) Iron and manganese oxidation and precipitation, and reduction and dissolution.

(d) Chromium chemical forms, adsorption, oxidation and reduction.

(e) Other metals adsorption and/or precipitation.

(f) Hydrogen sulfide chemical forms, adsorption, biological uptake, and stripping.

(g) Oxygen used in decomposition, produced by algae photosynthesis, and reaeration.

(h) Carbon inorganic and organic forms, and biological uptake.

(i) Methane may be formed from decaying vegetation under anaerobic conditions; can be stripped to atmosphere over spillway or by passage through dam and/or turbines.

(5) Bacterial die-away which is a function of temperature and residence time.

(6) Gas or nitrogen supersaturation due to air entrainment in stilling basins; high nitrogen gas concentration can cause fish impacts.

Another basic issue related to water quality impacts of reservoirs involves the consideration of reservoir input loadings of different potential pollutants. Key water pollutant sources and consequent loadings include the following:

(1) Point source discharges upstream of the reservoir, these can include treated and/or untreated discharges from urban areas and industrial activities.

(2) Nonpoint source discharges upstream of the reservoir; these include runoff from urban, industrial, agricultural, and undeveloped areas; some key pollutants include sediment, nutrients, organics, and pesticides.

(3) In-reservoir decomposition of terrestrial flora left in place during reservoir filling; anaerobic conditions near the bottom or in the hypolimnion can cause problems with methane and hydrogen sulfide formation and then their subsequent release.

(4) Uses of water in the reservoir and resultant quality changes; could be concerns regarding organics, nutrients, and bacteria.

(5) Direct precipitation and/or dry deposition into reservoir; nutrients and metals as well as acid rain may be of concern.

(6) Ground water flow into reservoir.

Finally in terms of background considerations, it is pertinent to consider which water quality parameters should be modeled. Typical parameters or characteristics subject to modeling include:

(1) Dissolved oxygen and biochemical oxygen demand.

(2) Nutrients.

(3) Temperature.

(4) Others (pH, iron, solids, pesticides, fecal coliforms, methane, hydrogen sulfide, etc.).

(5) "Waste assimilative capacity" of the reservoir.

8.9 Timing of Water Quality Modeling

Water quality modeling should be done during several phases over the project life. Different purposes may be associated with modeling for a given phase. The phases can be considered in terms of pre-construction planning, construction, transition years, and continued operation.

8.10 Pre-construction Planning Phase

Modeling should begin prior to the finalization of the reservoir design and the initiation of the construction phase. This modeling would typically be included in an environmental impact statement. The purpose should be to predict potential problems and then develop mitigation measures to minimize these problems. The modeling should anticipatorily address the construction phase, transition years, and continued operation of the reservoir.

8.11 Construction Phase

Real time modeling should be done during this phase, with several-fold purposes to be achieved. Several years might be required for construction, and most of the terrestrial flora in the inundated area is typically left in place and is in various states of decomposition. A water quality monitoring program should be conducted during this phase. The monitoring program would allow the calibration/verification of the water quality models used during this phase. In addition, the modeling should facilitate the implementation of positive environmental management programs. Examples of such programs include the planning of specific construction activities for specific seasons, and adjustment in selective removal of terrestrial flora in the area being inundated.

8.12 Transition Years Phase

Real-time modeling should be done during the transition period between the completion of the filling of the reservoir and a point in time when the reservoir water quality exhibits repeatable patterns.

This period would typically encompass the time of the most rapid decomposition of the terrestrial flora in the inundated area. While no specific time can be identified, periods in the order of 5 to 10 years might be a good estimate. Again, it is assumed that a water quality monitoring program is in place. As noted earlier, the monitoring program would allow the calibration/verification of the utilized water quality models. The modeling could also be used to evaluate and facilitate the implementation of positive environmental management programs. Examples of such program include the use of artificial aeration at selected locations within the reservoir, and the use of water level drawdown to expose and allow the selective removal of decaying terrestrial flora from the fringes of the inundated area.

8.13 Continued Operation Phase

This phase represents the long-term operation of the reservoir. It could extend for several decades and even beyond. Water quality modeling during this phase is important in that longer time periods are involved and greater opportunities exist for the planning and implementation of positive water quality management programs. Monitoring could continue to be used for calibration/verification of utilized water quality models. It is assumed that more sophisticated models would be selected/developed/used during this phase. Potential water quality impact mitigation measures which could be used during this phase include the design and use of in reservoir artificial aeration measures, and the planning and implementation of land use controls and watershed management measures.

Several conclusions can be drawn from this brief review of water quality modeling for tropical reservoirs. These conclusions include:

(1) The purposes of water quality modeling can include the development of an understanding of the lake and reservoir processes; and the development of information which can be used to plan, design and manage environmentally the project.

(2) A selection process should be used for models so as to match the model with data availability and phase of study. The assumptions and limitations of any selected model should be considered in interpreting the results of modeling efforts.

(3) The most effective modeling requires a long-term commitment to both modeling and monitoring efforts.

(4) Modeling can be used to establish linkages to watershed management, operational controls, and downstream developments.

References

Cole, T.M. and Wells, S.A., 2002, CE-QUAL-W2: A two-dimensional, laterally averaged, hydrodynamic and water quality model, version 3.1,: *U.S. Army Corps of Engineers Instruction Report* EL-02.

Rounds, S.A., Wood, T.M., and Lynch, D.D., 1999, Modeling discharge, temperature, and water quality in the Tualatin River, Oregon: *U.S. Geological Survey Water-Supply Paper* 2465-B, pp.121.

Rounds, S.A. and Wood, T.M., 2001, Modeling water quality in the Tualatin River, Oregon, 1991-1997: *U.S. Geological Survey Water-Resources Investigations Report* 01-4041, pp.53.

❑❑❑

Chapter 9

Modeling: Oceans and Seas

9.1. Introduction

Oceans (from Okeanos in Greek) are saline waters that cover almost three quarters (71%) of the surface of the Earth. The area of the oceans is 361 million sq. km., and nearly half of the world's marine waters are over 3,000 meters deep.

Though somewhat arbitrarily divided into several "separate" oceans, these oceans are in fact one global, interconnected body of salt water, often called the World Ocean. The major divisions are defined in part by the continents and a variety of archipelagos, and are labeled the Pacific Ocean, the Atlantic Ocean, the Indian Ocean, the Southern Ocean, and the Arctic Ocean. Smaller regions of the oceans are called seas, gulfs and straits.

There are a some smaller bodies of salt water that are not interconnected with the World Ocean. These are not considered to be oceans or parts of oceans, though some of them have been given the name, sea.

Geologically, an ocean is an area of oceanic crust covered by water. Oceanic crust is the thin layer of solidified volcanic basalt that covers the Earth's mantle where there are no continents. From this point of view, there are three "oceans" today: the World Ocean, and the Black and Caspian Seas that were formed by the collision of Cimmeria with Laurasia. The Mediterranean Sea is very nearly its own "ocean", being connected to the World Ocean through the Strait of Gibraltar, and indeed several times over the last few million years movement of the African Continent has closed the strait off entirely, making the Mediterranean a fourth "ocean".

9.2. Physical Properties

The area of the World Ocean is 361 million square kilometers, its volume is over 1,340 million cubic kilometers, and its average depth is 3,711 meters. Nearly half of the world's marine waters are over 3,000 meters deep. The vast abyssal plains of the deep ocean cover about 40% of the Earth's surface. This does not include seas not connected to the World Ocean, such as the Caspian Sea. The total mass of the hydrosphere is about 1.4×10^{21} kilograms, which is about 0.023% of the Earth's total mass (Wikipedia, 2008).

9.3 Exploration

Travel on the surface of the ocean through the use of boats dates back to prehistoric times, but only in modern times has extensive underwater travel become possible.

The deepest point in the ocean is the Marianas Trench located in the Pacific Ocean near the Northern Mariana Islands. It has a maximum depth of 10,923 meters. It was fully surveyed in 1951 by the British naval vessel, "Challenger II" which gave its name to the deepest part of the trench, the "Challenger Deep".

Much of the bottom of the world's oceans are unexplored and unmapped. A global image of many underwater features larger than 10 kilometers is created in 1995 based on gravitational distortions of the nearby sea surface.

9.4 Regions

Oceans are divided into numerous regions depending on the physical and biological conditions of these areas. The pelagic zone includes all open ocean regions, and can be subdivided into further regions categorised by depth and light abundance. The photic zone covers the oceans from surface level to 200 metres down. This is the region where the photosynthesis most commonly occurs and therefore contains the largest biodiversity in the ocean. Since plants can only survive with photosynthesis any life found lower than this must either rely on material floating down from above or find another primary source; this often comes in the form of hydrothermal vents in what is known as the aphotic zone. The pelagic part of the photic zone is known as the epipelagic (Fig. 9.1).

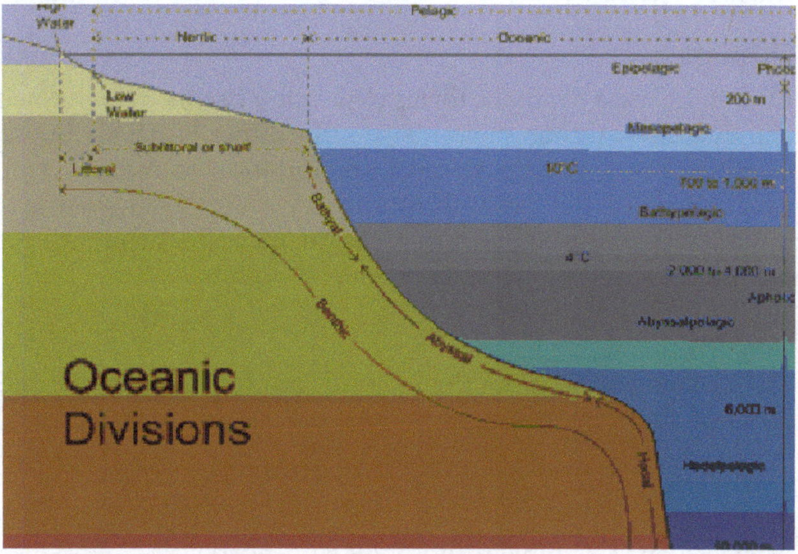

Fig. 9.1 The Major Oceanic Divisions

The pelagic part of the aphotic zone can be further divided into regions that succeed each other vertically. The mesopelagic is the uppermost region, with its lowermost boundary at a thermocline of 10°C, which, in the tropics generally lies between 700 and 1,000m. After that is the bathypelagic lying between 10°C and 4°C, or between 700 or 1,000m and 2,000 or 4,000m. Lying along the top of the abyssal plain is the abyssalpelagic, whos lower boundary lies at about 6,000m. The final zone falls into the oceanic trenches, and is known as the hadalpelagic. This lies between 6,000m and 10,000m and is the deepest oceanic zone.

Along with pelagic aphotic zones there are also benthic aphotic zones, these correspond to the three deepest zones. The bathyal zone covers the continental slope and the rise down to about 4,000m. The abyssal zone covers the abyssal plains between 4,000 and 6,000m. Lastly, the hadal zone corresponds to the hadalpelagic zone which is found in the oceanic trenches. The pelagic zone can also be split into two subregions, the neritic zone and the oceanic zone.

The neritic encompasses the water mass directly above the continental shelves, while the oceanic zone includes all the completely open water. In contrast, the littoral zone covers the region between low and high tide and represents the transitional area between marine and terrestrial conditions. It is also known as the intertidal zone because it is the area where tide level affects the conditions of the region.

9.5 Climate

One of the most dramatic forms of weather occurs over the oceans: tropical cyclones. Ocean currents greatly affect Earth's climate by transferring warm or cold air and precipitation to coastal regions, where they may be carried inland by winds. The Antarctic Circumpolar current encircles that continent, influencing the area's climate and connecting currents in several oceans.

9.6 Ecology

The oceans are home to the majority of plant and animal life on Earth. These lifeforms include:

- Radiata
- Fish
- Cetacea such as whales, dolphins and porpoises,
- Cephalopods such as the octopus
- Crustaceans such as lobsters and shrimp
- Marine worms
- Plankton (Phyto and Zoo)
- Krill

9.7 Economy

The oceans are essential for transportation of goods and passengers: most of the world's goods are moved by ship between the countries sea ports from Singapore port at Singapore to Mangalore port in India. Important ship canals include the Saint Lawrence Seaway, Panama Canal, and Suez Canal. They are also an important source of valuable foodstuffs via the fishing industry.

9.8 Ancient Oceans

Continental drift has reconfigured the Earth's oceans, joining and splitting ancient oceans to form the current oceans. Ancient oceans include:

- Bridge River Ocean, the ocean between the ancient Insular Islands and North Amarica.
- Iapetus Ocean, the southern hemisphere ocean between Baltica and Avalonia.
- Panthalassa, the vast world ocean that surrounded the Pangaea supercontinent.
- Rheic Ocean
- Slide Mountain Ocean, the ocean between the ancient Intermontane Islands and North America.
- Tethys Ocean, the ocean between the ancient continents of Gondwana and Laurasia.

9.9 Tide

Tides are the cyclic rising and falling of Earth's ocean surface caused by the tidal forces of the Moon and the Sun acting on the Earth. Tides cause changes in the depth of the sea, and also produce oscillating currents known as tidal streams, making prediction of tides important for coastal navigation. The strip of seashore that is submerged at high tide and exposed at low tide, the intertidal zone, is an important ecological product of ocean tides.

The changing tide produced at a given location on the Earth is the result of the changing positions of the Moon and Sun relative to the Earth coupled with the effects of the rotation of the Earth and the local bathymetry. Though the gravitational force exerted by the Sun on the Earth is almost 200 times stronger than that exerted by the Moon, the tidal force produced by the Moon is about twice as strong as that produced by the Sun. The reason for this is that the tidal force is related not to the strength of a gravitational field, but to its gradient. The field gradient decreases with distance from the source more rapidly than does the field strength; as the Sun is about 400 times further from the Earth than is the Moon, the gradient of the Sun's field, and thus the tidal force produced by the Sun, is weaker (Fig. 9.2).

9.9.1 Tidal terminology

The maximum water level is called "high tide" or "high water" and the minimum level is "low tide" or "low water." If the ocean were a constant depth, and there were no land, high water would occur as two bulges in the height of the oceans one bulge facing the Moon and the other on the opposite side of the earth, facing away from the Moon. There would also be smaller, superimposed bulges on the sides facing toward and away from the Sun. At any given point in the ocean, there are normally two high tides and two low tides each day just as there would be for an earth with no land; however, rather than two large bulges propagating around the earth, with land masses in the way the result is many smaller bulges propagating around amphidromic points, so there is no simple, general rule for predicting the time of high tide from the position of the Moon in the sky. The common names of the two high tides are the "high high" tide and the "low high" tide; the difference in height between the two is known as the "daily inequality." The daily inequality is generally small when the Moon is over the equator. The two low tides are called the "high low" tide and the "low low" tide. On average, high tides occur 12 hours 24 minutes apart. The 12 hours is due to the Earth's rotation, and the 24 minutes to the Moon's orbit. This is the "principal lunar semi-diurnal" period, abbreviated as the M2 tidal component, and it is, on average, half the time separating one lunar zenith from the next. The M2 component is usually the biggest one, but there are many others as well due to such complications as the tilt of the Earth's rotation axis and the inclination of the lunar orbit. The lunar cycle is what is tracked by tide clocks.

The time between high tide and low tide, when the water level is falling, is called the "ebb." The time between low tide and high tide when the tide is rising, is called "flow," or "flood." At the times of high tide and low tide, the tide is said to be "turning," also slack tide. The height of the high and low tides also varies. Around new and full moon when the Sun, Moon and Earth form a line, the tidal forces due to the Sun reinforce those of the Moon. The tides' range is then at its maximum: this is called the "spring tide," or just "springs" and is derived not from the season of spring but rather from the verb "to jump" or "to leap up." When the Moon is at first quarter or third quarter, the Sun and Moon are at 90° to each other and the forces due to the Sun partially cancel out those of the Moon. At these points in the lunar cycle, the tide's range is at its minimum: this is called the "neap tide," or "neaps".

Spring tides result in high waters that are higher than average, low waters that are lower than average, slack water time that is shorter than average and stronger tidal currents than average. Neaps result in less extreme tidal conditions. Normally there is a seven day interval between springs and neaps.

The relative distance of the Moon from the Earth also affects tide heights: When the Moon is at perigee the range increases, and when it is at apogee the range is reduced. Every 7½ lunations, perigee and either a new or full moon coincide; at these times the range of tide heights is greatest of all, and if a storm happens to be moving onshore at this time, the consequences can be especially severe. The effect is enhanced even further if the line-up of the Sun, Earth and Moon is so exact that a solar or lunar eclipse occurs concomitant with perigee.

9.9.2 Timing

In most places there is a delay between the phases of the Moon and its effect on the tide. Springs and neaps in the North Sea, for example, are two days behind the new/full Moon and first/third quarter, respectively. The reason for this is that the tide originates in the southern oceans, the only place on the globe where a circumventing wave can travel unimpeded by land.

The resulting effect on the amplitude, or height, of the tide travels across the oceans. It is known that it travels as a single broad wave pulse northwards over the Atlantic. This causes relatively low tidal ranges in some locations and high ones in other places. This is not to be confused with tidal ranges caused by local geography, as can be found in Nova Scotia, the Bristol Channel, the Channel Islands, and the English Channel. In these places tidal ranges can be over 10 metres.

The Atlantic tidal wave arrives after approximately a day in the English Channel area of the European coast and needs another day to go around the British Isles in order to have an effect in the North Sea. Peaks and lows of the Channel wave and North Sea wave meet in the Strait of Dover at about the same time but generally favour a current in the direction of the North Sea.

The exact time and height of the tide at a particular coastal point is also greatly influenced by the local bathymetry. There are some extreme cases: the Bay of Fundy, on the east coast of Canada, features the largest well-documented tidal ranges in the world, 16 metres, because of the shape of the bay. Southampton in the United Kingdom has a double high tide caused by the flow of water around the Isle of Wight, and Weymouth, Dorset has a double low tide because of the Isle of Portland. Ungava Bay in Northern Quebec, north eastern

Canada, is believed by some experts to have higher tidal ranges than the Bay of Fundy, but it is free of pack ice for only about four months every year, whereas the Bay of Fundy rarely freezes even in the winter.

9.9.3 Tidal physics

Ignoring complications arising from ocean currents, the ocean's surface is closely approximated by an equipotential surface, which is commonly referred to as the geoid. Since the gravitational force is equal to the gradient of the potential, there are no tangential forces on such a surface, and the ocean surface is thus in gravitational equilibrium.

The Moon exerts its gravitational pull differently on different parts of the earth. The farther the Moon, the weaker its pull. Imagine a shell of the outer Earth, this diagram shows the Moon's gravity differential over the thickness of the shell.

The Moon's gravity differential field at the surface of the earth is known as the Tide Generating Force. This is the primary mechanism that drives tidal action and explains two bulges, accounting for two high tides per day. Other forces, such as the Sun's gravity, also add to tidal action (Fig 9.2).

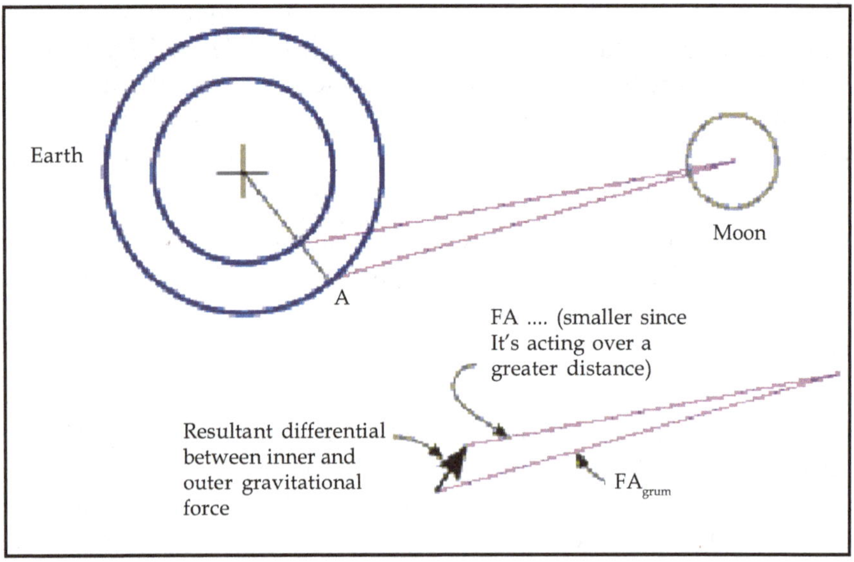

Fig. 9.2: Forces Acting on Earth and Moon

For simplicity, consider the tidal influence cause by the Moon: it should be understood that the same desciption applies to the Sun as well. At the point right "under" the Moon, the water is closer than the solid Earth; so it is pulled more and rises. On the opposite side of the Earth, facing away from the Moon, the water is farther from the moon than the solid earth, so it is pulled less and effectively moves away from Earth, rising as well. On the lateral sides, the water is pulled in a slightly different direction than at the centre. The vectorial difference with the force at the centre points almost straight inwards to Earth. It can be shown that the forces at the sub-lunar and antipodal points are approximately equal and that the inward forces at the sides are about half that size. Somewhere in between there is a point where the tidal force is parallel to the Earth's surface (Fig. 9.3).

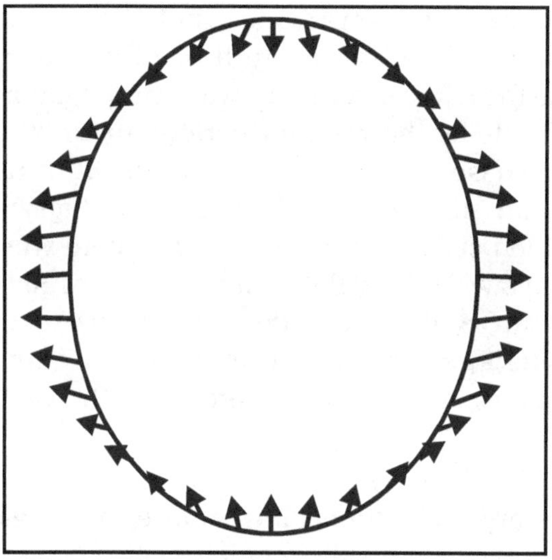

Fig. 9.3 Tidal Force On Earth

Those parallel components actually contribute most to the formation of tides, since the water particles are free to follow. The actual force on a particle is only about a ten millionth of the force caused by the Earth's gravity.

9.9.4 Tidal amplitude and cycle time

Since the Earth rotates relative to the Moon in one lunar day, each of the two tidal bulges travels around at that speed relative to an observer on Earth, leading to one high tide every 12 hours and 24 minutes. The theoretical amplitude of oceanic tides due to the Moon is about 54 cm at the highest point, which corresponds to the amplitude that would be reached if the ocean possessed a uniform depth, there were no landmasses, and the Earth was not rotating. The Sun similarly causes tides, of which the theoretical amplitude is about 25 cm with a cycle time of 12 hours. At spring tide the two effects add to each other to a theoretical level of 79 cm, while at neap tide the theoretical level is reduced to 29 cm. Since the orbit of the Earth about the Sun, and the Moon about the Earth, are elliptical, the amplitudes of the tides change somewhat as a result of the varying Earth-Sun and Earth-Moon distances. This causes a variation in the tidal force and theoretical amplitude of about ±18% for the Moon and ± 5% for the Sun. If both the Sun and Moon were at their closest positions and aligned at new moon, the theoretical amplitude would reach 93 cm.

Real amplitudes differ considerably, not only because of variations in ocean depth, and the obstacles to flow caused by the continents, but also because the natural period of wave propogation is of the same order of magnitude as the rotation period: about 30 hours. If there were no land masses, it would take about 30 hours for a long wavelength ocean surface wave to propagate halfway around the Earth. If the Moon suddenly vanished, and there were no land, the level of the oceans would oscillate with a period of 30 hours and with an amplitude that would slowly decrease as energy was dissipated. This 30 hour value is a simple function of terrestrial gravity, the average depth of the oceans, and the circumference of the Earth.

9.9.5 *Tides and Navigation*

Tidal flows are of profound importance in navigation and very significant errors in position will occur if they are not taken into account. Tidal heights are also very important; for example many rivers and harbours have a shallow "bar" at the entrance which will prevent boats with significant draught from entering at certain states of the tide.

The timings and velocities of tidal flow can be found by looking at a tidal chart or tidal stream atlas for the area of interest. Tidal charts come in sets, with each diagram of the set covering a single hour between one high tide and another and give the average tidal flow for that one hour. An arrow on the tidal chart indicates the direction and the average flow speed for spring and neap tides. If a tidal chart is not available, most nautical charts have "tidal diamonds" which relate specific points on the chart to a table of data giving direction and speed of tidal flow.

Standard procedure to counteract the effects of tides on navigation is to (1) calculate a "dead reckoning" position (or DR) from distance and direction of travel, (2) mark this on the chart (with a vertical cross like a plus sign) and (3) draw a line from the DR in the direction of the tide. The distance the tide will have moved the boat along this line is computed by the tidal speed, and this gives an "Estimated Position" or EP.

Nautical charts display the "charted depth" of the water at specific locations by the use of contour plots. These depths are relative to a "chart datum", which is the level of water at the lowest possible astronomical tide and are therefore the minimum water depth possible during the tidal cycle. "Drying heights" may also be shown on the chart, which are the heights of the exposed seabed at the lowest astronomical tide.

Heights and times of low and high tide on each day are published in tide tables. The actual depth of water at the given points at high or low water can easily be calculated by adding the charted depth to the published height of the tide. The water depth for times other than high or low water can be derived from tidal curves published for major ports. If an accurate curve is not available, the rule of twelfths can be used. This approximation works on the basis that the increase in depth in the six hours between low and high tide will follow this simple rule: first hour - 1/12, second - 2/12, third - 3/12, fourth - 3/12, fifth - 2/12, sixth - 1/12.

9.9.6 Other tides

In addition to oceanic tides, there are atmospheric tides as well as terrestrial tides that affect the rocky mass of the Earth. Atmospheric tides may be negligible for everyday phenomena, drowned by the much more important effects of weather and solar thermal tides. However, there is no strict upper limit to the Earth's atmosphere, and the tidal pull increases with the distance from the Earth's centre.

The Earth's crust, on the other hand, rises and falls imperceptibly in response to the Moon's gravitational attraction. The amplitude of terrestrial tides can reach about 55 cm at the equator, and they are nearly in phase with the Moon.

While negligible for most human activities, terrestrial tides need to be taken in account in the case of some particle physics experimental equipments. Among the effects that need to be taken into account are circumference deformation for circular accelerators and particle beam energy.

Since tidal forces generate currents of conducting fluids within the interior of the Earth, they affect in turn the Earth's magnetic field itself. Tsunamis, the large waves that occur after earthquakes, are sometimes called tidal waves, but have nothing to do with the tides. Other phenomena unrelated to tides but using the word tide are rip tide, storm tide, hurricane tide, and red tide.

9.10 Ocean Modeling

To a large extent the dynamic state of the ocean can be described by the velocity, temperature, salinity and pressure at each point within it. In ocean models, temperature is usually represented in the form of potential temperature with relative to a pressure of one atmosphere because this remains constant under adiabatic changes in pressure. Salinity is a composite variable defined by an international standard which represents the combined effect of the different dissolved salts in the ocean (Fig. 9.4).

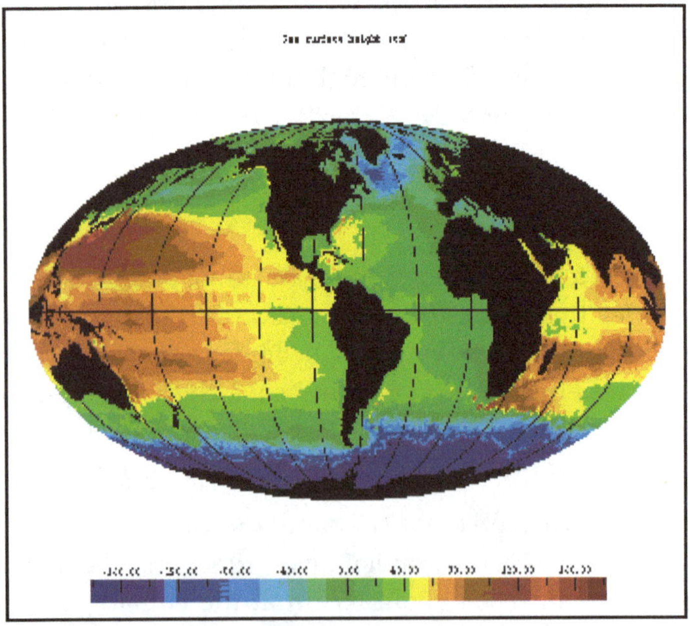

Fig. 9.4 Ocean Surface Heights with Different Gradients

The evolution of such a system can be described by a three dimensional momentum equation, an advection-diffusion equation for heat and salt, a continuity equation and an equation of state which defines the density in terms of the temperature, salinity and pressure.

In ocean models, three important approximations are made to reduce the computational load. The first is to assume, in the continuity equation, that the ocean is incompressible. The second is to assume, in the vertical momentum equation, that the vertical velocity is small and that the terms involving it can be neglected. The third is to assume, in the horizontal momentum equations, that small changes in density can be neglected except where they affect the horizontal pressure gradient. (Bryan 1969).

The main prognostic variables are u the horizontal velocity, T the potential temperature and S the salinity. The other variables, p the pressure, r the density and w the vertical velocity can be calculated from the prognostic variables. In these equations t is time and f is the Coriolis term (equal to 2 omega sin (theta), where omega is the Earth's rotation rate and theta is the latitude. g represents gravity. The terms Du, Dt and Ds represent diffusion of u, T and S respectively. The terms Fu, Ft and Fs represent the forcing of u, T and S respectively.

The horizontal velocity u is zero on solid sidewall boundaries. The gradients of potential temperature and salinity normal to all solid boundaries (including the bottom) are also zero. The upper surface boundary conditions are used to specify the exchange of heat and fresh water at the air-sea interface and the stress acting on the ocean due to the wind.

Bryan (1969) developed a computer scheme for solving these equations which has been widely followed. This splits the ocean up into a large number of boxes along lines of constant latitude, longitude and depth. In a coarse resolution ocean model the individual boxes will have a width of typically 2 degrees in the horizontal and a thickness ranging from 20 m for layers near the ocean surface to 500 m at depth in the ocean. Topography is represented by assuming that each box is either a sea box or a land box.

The state of the ocean is represented by values of potential temperature, salinity and horizontal velocity at the centre of each box. The above equations are then integrated over each box to give an equation in which the advection and diffusive terms are replaced by the fluxes through the boundaries of the box and the other terms written in terms of averages over the box.

The equations are then cast in finite difference form with gradients calculated from values at neighbouring boxes. For accuracy the Arakawa-B grid is used, in which the velocities are specified at points offset from those at which tracers are specified. This scheme is better than the other standard Arakawa schemes at representing the propagation of oceanic Rossby waves, when the grid spacing is greater than the Rossby radius of the ocean. The Rossby radius is typically 25 km for the first internal mode of the ocean but is much smaller for higher modes. Even with high resolution ocean models, using grids of 20 km or less, the Arakawa-B grid is still preferred because of its better performance in representing the higher vertical modes.

The equations are used to step the model solution forward in time in a series of discrete timesteps. This is computationally most efficient when the timestep is as large as possible. However for stability reasons the timestep has to be less than the time it takes a wave to move from one grid box to the next.

The fastest wave in the ocean, the external gravity wave, has a speed of about 250 m/s. The next fastest wave, the first internal gravity wave, has a speed of about one hundredth of this value. If the external gravity wave, which has little effect on the ocean circulation, is filtered out it enables the timestep to be increased by a factor of 100, greatly speeding up the model.

Bryan (1969) did this by placing a rigid lid on the ocean. The lid removes the external gravity waves but he was able to show that it had little effect on the large scale circulation of the model ocean. The approximation introduces an additional stream function equation which has to be solved but the extra computer time required to do this is relatively small.

Bryan's scheme uses a leapfrog timestep to step the equations forward in time. This is computationally efficient and propagates waves correctly without any change in amplitude. However the scheme is unstable when used with diffusive terms and so for these a simple Euler forward timestep is used. The leapfrog scheme can lead to a splitting of the solution on even and odd timesteps so to prevent this growing too large a forward timestep is introduced every 20 to 100 timesteps for all the terms. This can use the Euler forward scheme but usually a Euler backward scheme is used because this does not increase the amplitude of any waves present, an unphysical process which can readily lead to numerical instability. Fuller details about the model and the numerical schemes used are given by Bryan (1969), Semtner (1973) and Cox (1984). Good reviews of the properties of the different finite difference schemes are given by Mesinger and Arakawa (1976) and by Roach (1976).

9.11 Development of the Model

Later developments of the model are based primarily on Semtner's (1974) revision of the initial code. He added Takano's (1974) scheme for handling islands and streamlined the code so that it ran efficiently on vector processing computers.

Ocean models are computationally very expensive to run and so should make very efficient use of the computer time available. For a large ocean model this requires the inner computational loops to be very efficient. Semtner introduced a large number of small vector loops which ran very quickly on the early vector processing machines. He also introduced the idea of moving data between main memory and disk storage in slabs, each containing all the data from one latitude band of the model. This allowed the model to be vectorised in the horizontal, usually giving vectors of 90 elements or more, so that the start up overhead in each vector operation was small. By allowing slabs to be pre fetched it also meant that the central processing unit is only rarely left waiting for input-output operations to complete.

The program is organised with a main program which handles the beginning and end of the run and which includes the main time stepping loop. The details of each time step are handled by routine 'step'. This calls subroutine 'clinic' to calculate the new baroclinic velocity fields, 'tracer' to calculate the new fields of potential temperature and salinity, and finally calls routine 'relax' to solve the stream function equation.

9.12 The Cox Code

Semtner's code is widely used by the oceanographic community but it was eventually superseded by the code developed by Cox (1984). This introduced a large number of new options, for example different vertical mixing schemes, and it allowed the horizontal grid spacing to be varied.

This version of the code is also written so that each slab is treated by the processor as a single long vector. This scheme is introduced for use on the Cyber 205 computer, whose vector processing unit is very fast but had a long start up time. Unfortunately, because the scheme carries out many redundant calculations over land points, it is less efficient than the Semtner code on most other computers, be they workstations or Cray supercomputers. The Cox code also replaces some small arrays used in the Semtner code by arrays which are the size of a full slab. These increase the number of main memory to cpu transfers and so will slow down most other computers.

9.13 The MOM Code

More recently the group at GFDL has revised the Cox code and added a large number of useful options. The new code is called the GFDL Modular Ocean Model or MOM code (Pacanowski *et al,* 1990). Options and common block specifications are now included using Unix 'C' compiler preprocessing directives. The GFDL group has also removed the very large vectors of the Cox version, so that the code works efficiently on most modern computers.

Replacing the stream function equation by a pressure equation The treatment of the island boundary conditions in the stream function routine involves an integration around the perimeter of each island. This can be done efficiently on a computer with scalar architecture but this is not so on vector or array processor computers.

To overcome this problem Smith *et al* (1992) recently proposed replacing the stream function equation by an equation for the surface pressure field. The method for solving the pressure equation is similar to that used for the stream function equation but the island boundary condition no longer involves an integral. As a result the new method works with equal efficiency on scalar, vector and array processor computers. For problems with no islands it requires a similar amount of computer time as the stream function scheme.

9.14 The Free Surface Code

Because of the limits in computer power, most early models of the ocean were of modest size, with typically less than 100 points in the horizontal. For such models the stream function scheme is the one most efficient for timestepping the barotropic velocity field. In a rectangular ocean a fast fourier transform method can be used to solve the equation, but in practice, because the boundary of the ocean is very complex, a relaxation scheme has to be used. This iterates the solution each timestep, the number of iterations being the same order as the number of grid points in the horizontal.

With large models, an alternative scheme is to timestep the baroclinic velocity field separately, using a vertically integrated velocity equation. This is conventionally called a 'free surface model' because it also removes the rigid lid constraint and includes a variable representing the surface elevation of the ocean. The scheme requires approximately 100 small timesteps for each timestep of the baroclinic velocity field and the tracer fields. Because the number of timesteps does not increase as the size of the model increases, the scheme is eventually more efficient than the stream function or pressure equation methods. A free surface model for use with the Cox code was developed by Killworth *et al* (1989).

9.15 The Array Processor Ocean Model

The reason for developing the present model arose out of a decision by the UK ABRC committee to buy a large array processing computer in 1993/94 for the UK research community. All the ocean model codes described above since Semtner have been designed for efficient use on fast vector processing computers. However the development of vector processors is starting to slow down and in future the main advances in computer power are expected to arise from the development of large arrays of small fast computer units each with its own internal storage. The units will be able to exchange data with each other, but this will be at a rate much slower than the internal transfer rate.

The way that we imagine such an array processor will be used to run an ocean model is that the ocean will be split up into small volumes each handled by a different processor. The processors will carry out time steps in parallel, each processor requesting boundary data from its 'adjacent' processors when required.

We expect that the most efficient scheme will be the one which chooses the volumes to mimimise the maximum amount of computing carried out by any of the processors and which minimises the maximum amount of data being transferred between one processor and its neighbours. These constraints will probably produce compact regions containing roughly equal number of oceanic grid points.

An exact minimum is probably not important, so in the code presented here it is assumed that the cells constituting each volume form a compact group in the horizontal and within the region contain all the cells extending from the sea surface to the bottom. Data that needs to be fetched from another processor will always be for cells at the sides of those being processed and never for cells above or below.

This assumption allows the model code to be vectorised in the vertical direction. The way that the ocean region is partitioned between the processors in the horizontal is left to the programmer or compiler to choose, depending on the geometry of the ocean basin being studied. For simplicity, all the do loops involving horizontal indices have been collected in one high level subroutine, so only this routine will need to be changed.

9.16 The Free Surface Model

Because the mom code is intended for use with large ocean models it uses a free surface model. The free surface subroutine is based on that of Killworth *et al* (1989) and uses the Euler backward timestepping option because this is efficient at damping out high frequency surface gravity waves. This prevents aliasing problems associated with the long baroclinic and short barotropic time steps.

The scheme used differs from that of Killworth *et al* in that the frictional terms are no longer included within the free surface subroutine. Instead they are calculated once only within each baroclinic timestep. This modification is computationally more efficient and has very little effect on the solution.

A stream function version of the model is also available. It was used to validate the free surface code but contains a stream function solving routine which has not yet been optimised for array processor computers. The stream function version is discussed further in the section on scalar and vector computers.

9.17 Advection in the Baroclinic Velocity Equation

In the primitive equation system of equations, the vertical velocity is required at two places, once to advect tracer quantities in the vertical, and once to transport momentum in the vertical. Using the Arakawa B grid these are required at different positions, so in the standard primitive equation model different schemes are used to calculate the two sets of vertical velocities.

Experience with the model has shown that the vertical velocity field calculated at tracer points is well behaved, but the vertical velocities calculated at velocity points is very noisy. This may be due to the way horizontal advection is defined at velocity points.

When solving the velocity equations, the flux of momentum through the side face of a grid box is calculated from the two velocity points on either side of the face. The divergence of this flow is then used to calculate the fluxes through the top and bottom faces of the box. When this scheme was checked within the free surface code, it was found that purely horizontal flow fields could result in significant spurious vertical velocities. Such velocities did not occur in the equations for tracers. A check against the standard stream function code showed that the barotropic velocities, calculated from the stream function, produced no spurious vertical velocities but the baroclinic flow field still did so.

The simplest scheme found which overcame these properties defines the advective flux through each side of a velocity box to be the average of the corresponding fluxes for the four surrounding tracer boxes. The new scheme is used in the present free surface version of the code and is recommended for use with stream function version of the code. The suggestion to use a vertical velocity field at velocity boxes which is the average of that at the surrounding tracer boxes has been made previously. However for consistency, the modified horizontal advection scheme needs to be used as well.

9.18 Validations

Validation tests are carried out using the old form of the horizontal advection code at velocity points. Initially the stream function version of the model was validated by checking against an equivalent test model using the MOM code. The only differences found were at the rounding error level.

The free surface version of the code is then validated against the stream function version with the barotropic valocity set to zero. This essentially validated those parts of the code concerned with solving the temperature, salinity and baroclinic velocity equations. Again the only differences are at the rounding error level.

The free surface code itself is checked against the stream function version by running the models with only the free surface and barotropic velocity fields switched on. The agreement between the two solutions is good, the kinetic energy of the free surface solution being just a few per cent higher. It also showes some long term oscillations. Both differences are presumably due to gravity and Kelvin waves present in the free surface solution but filtered out of the stream function solution.

Finally the free surface version is checked against the stream function with all parts of the code switched on. Again there was good agreement between the two solutions.

9.19 Schemes for Partitioning the Ocean

As discussed earlier, the present code makes no assumptions about the way the ocean will be partitioned between the individual array processors but for convenience places all the relevant loops over the horizontal indices in subroutine 'step'. In practice a number of schemes could be used.

The simplest, and the one likely to be used by most compilers, is to split the ranges of the latitude and longitude indices equally between the different processors. Each processor is then responsible for storing all the variables and for carrying out all the calculations for its own rectangular region of ocean. When processing points on its own boundary it sends requests to the neighbouring processor for any data it needs for grid points in the neighbouring region. Ocean resources are numerous from elements to aquatic animals. It should be conserved with appropriate technology and engineering principles with help of water quality modeling studies.

References

Bryan, K., 1969: A numerical method for the study of the circulation of the circulation of the world ocean. *Journal of Computational Physics*, 4, 347.

Cox, M.D., 1984: A primitive equation, 3-dimensional model of the ocean. GFDL *Ocean Group Technical Report No. 1*, Geophysical Fluid Dynamics Laboratory/NOAA, Princeton University, Princeton, N.J. 08542, U.S.A.

Killworth, P.D., Stainforth, D., Webb, D.J. & Paterson, S.M. 1989: A free-surface Bryan-Cox-Semtner ocean model. *IOSDL Report* No. 270. Institute of Oceanographic Sciences Deacon Laboratory, Wormley, 184pp.

Killworth, P.D., Stainforth, D., Webb, D.J. and Paterson, S.M. 1991: The development of a free-surface Bryan-Cox-Semtner ocean model. *Journal of Physical Oceanography*, 21, 1333-1348.

Mesinger, F., and Arakawa, A., 1976: Numerical methods used in atmospheric models. *GARP Publication series*, No. 17. World Meteorological Organization, 64pp.

Pacanowski, R.C., Dixon, K. and Rosati, A., 1990: The GFDL Modular Ocean Model users guide, version 1.0. *GFDL Group Technical Report* No. 2.

Roach, P.J., 1976: Computational Fluid Dynamics. Hermosa Publishers, Albuquerque, N.M., 446pp.

Semtner, A,J., 1974: A general circulation model for the World Ocean. *Technical Report* No. 9, Department of Meteorology, University of California, Los Angeles, 99pp.

Smith, R.D., Dukowicz, J.K., and Malone, R.C., 1992: *Parallel ocean general circulation modelling. Physica* D, 60, 38-61.

Takano, K., 1974: A general circulation model of the world ocean. "Numerical Simulation of weather and Climate". *Technical Report* No. 8, Department of Meteorology, University of California, Los Angeles, 47pp.

Wikipedia, 2007. World Oceans.

❒❒❒

Chapter 10

Water Quality Modeling

BOD and DO

10.1 Introduction

Biochemical Oxygen Demand (BOD) is a chemical procedure for determining how fast biological organisms use up oxygen in a body of water. It is usually performed over a 5 day period at 20 C. It is used in water quality management and assessment, ecology and environmental science. BOD is not an accurate quantitative test, although it could be considered as an indication of the quality of a water source.

10.2 Typical BOD values

Most pristine rivers will have a 5-day BOD of less than 1 mg/l. Moderately polluted rivers may have a BOD value in the range of 2 to 8 mg/l. Municipal sewage that is efficiently treated by a three stage process would have a value of about 20 mg/l. Untreated sewage varies, but averages around 600 mg/l in India and as low as 200 mg/l in the U.S., or where there is severe groundwater or surface water infiltration. Slurry from dairy farms is around 8000 mg/l and silage liquor around 60000 mg/l.

BOD Level (in ppm)	Water Quality of water
1 - 2	Very Good-not much organic waste present
3 - 5	Moderately clean
6 - 9	Moderately polluted
10<	Highly polluted

BOD measures the rate of oxygen uptake by micro organisms in a sample of water at a fixed temperature (20°C) and over a given period of time in the dark. To ensure that all other conditions are equal, a very small amount of micro-organism seed is added to each sample being tested. This seed is typically generated by diluting activated sludge with de-ionised water. The test generally takes place over an elapsed period of five days, but other BOD tests are also used.

The BOD test is carried out by diluting the sample with deionised water saturated with oxygen, inoculating it with a fixed aliquot of seed, measuring the dissolved oxygen and sealing the sample. The sample is kept at 20 °C in the dark to prevent photosynthesis for five days, and the dissolved oxygen is measured again. The difference between the final DO and initial DO is the BOD. The apparent BOD for the control is subtracted from the control result to provide the corrected value.

The loss of dissolved oxygen in the sample, once corrections have been made for the degree of dilution, is called the BOD_5. In India, allylthiourea is also added at the start of the test to prevent oxidation of ammonia. Results from such tests are represented as BOT5and referred to as Carbonaceous BOD (CBOD) in India. Less frequently used is the Ultimate BOD (UBOD) test, in which DO is repeatedly measured by DO meter in the same specialised bottles until it has reached equilibrium.

BOD is similar in function to chemical oxygen demand (COD), in that both measure the amount of organic compounds in water. However, COD is less specific, since it measures everything that can be chemically oxidised, rather than just levels of biologically active organic matter.

BOD is used as a gauge of the effectiveness of wastewater treatment plants. Various commercial devices are available for its determination.

BOD can be calculated by:

- *Undiluted:* Initial DO - Final DO = BOD
- *Diluted:* (Initial DO - Final DO)- BOD of Seed) x Dilution Factor

10.3 History of the Use of BOD

The *Royal Commission on River Pollution,* which was established in 1865 and the formation of the *Royal Commission on Sewage Disposal* in 1898 led to the selection in 1908 of BOD_5 as the definitive test for organic pollution of rivers. Five days was chosen as an appropriate test period because this is supposedly the longest time that river water takes to travel from source to estuary in the U.K. In 1912, the commission also set a standard of 20 ppm BOD_5 as the maximum concentration permitted in sewage works discharging to rivers, provided that there was at least an 8:1 dilution available at dry weather flow. This was contained in the famous 20:30 (BOD:Suspended Solids) + full nitrification standard which was used as a yardstick in the U.K. up to the 1970s for sewage works effluent quality.

The carbonaceous biological oxygen demand is an expression of the waters organic matter content. That is to say the biodegradable part of the organic matter which gives rise to oxygen consumption. The organic matter content is measured by registering the oxygen consumed during the degradation for a period of 5 days. The BOD units are therefore gO2/m3. Degradation in the environment of the organic matter expressed as BOD gives rise to an equivalent consumption of oxygen. The BOD degradation terms will therefore be part of the oxygen balance. Degradation of BOD is also a source of nutrients since these are part of the organic matter.

Biological Oxygen Demand, BOD being products of the BOD degradation can be oxidised and give rise to an additional oxygen consumption. The oxygen consumption and the nutrient production of the BOD degradation have no direct influence on the BOD degradation and on the mass balance of BOD itself. However, the modeling of BOD is an interrelated part of the dissolved oxygen (DO) modeling and the BOD degradation stops if the water becomes anaerobic, i.e. DO = ZERO. The differential equations describing the BOD variations and the differential equation for oxygen are coupled and solved simultaneously. The set of two differential equations represents the most simple BOD-DO model. At a more complex level, the BOD-DO model can include the production of nutrients during degradation of organic matter as well as the processes changing the oxidation level of the nitrogen. The consequences of these changes in oxidation levels for the oxygen balance are also included. This is described in detail under Nutrients. At the most complex levels, three fractions of BOD are considered: dissolved BOD (BODd), suspended BOD (BODs) and deposited BOD (BODb).

Depending on the chosen model level a set of up to seven coupled differential equations are used: oxygen, three fractions of BOD, ammonia, nitrate and temperature. The actual number of differential equations depends on the chosen model level. If modelling of phosphorus and/or coliforms are chosen, up to four additional differential equations or components are included. Activation of Wetland processes will not require additional components. The BOD in treated wastewater will be dissolved and/or suspended. The distribution between these two fractions will depend on the type of treatment plant. The ratio of dissolved BOD to suspended BOD will typically be in the range 1:1.5 - 1:1 for mechanically treated wastewater and around 2:1 for wastewater treated additionally by chemical precipitation.

The degradation of organic matter can be described by first order kinetics. All three fractions of BOD will be subject to decay though possibly at different rates. At model levels 5 and 6 a first order decay rate has to be specified for each fraction. The degradation of organic matter is temperature dependent. An Arrhenius expression is used to describe the temperature dependence and Arrhenius temperature coefficients are specified for one Water Quality Reference three BOD fractions depending on the model level. The input degradation constants must be specified for conditions at 20 °C. In addition to the loss by decay, suspended BOD will be lost by deposition and resuspension from the bottom into the water is assumed to occur.

The deposition will stop at water velocities above a certain critical velocity for deposition (Ucrit) and resuspension will only take place at water velocities above this critical value. For model levels 5 and 6 resuspension will occur only when the BODbconcentration is above a certain critical value. This means that BOD will be resuspended only when there is some BOD at the bed to resuspend. At model levels 2 and 4, the resuspension is assumed to be a zero order process, whereas for model levels 5 and 6 a first order expression is used. In other words, resuspension of levels 2 and 4 will continue as long as the flow velocity is higher than Ucrit. No balance for BODsed is modelled when using levels 2 or 4. The model constants, which have to be specified for BOD, are summarised below:

Model levels 1-4

K3 = The degradation constant for organic matter (1/day)

.3 = Arrhenius temperature coefficient for decay of organic matter Model levels 2 and 4

K3 = The degradation constant for organic matter (1/day)

.3 = Arrhenius temperature coefficient for decay of organic matter

K5 = Deposition rate for BOD (m/day)

S1 = Resuspension rate ($g/m^2/day$)

Vcrit = Critical velocity for deposition (m/s)

Model levels 5 and 6

Kd3 = The degradation constant for dissolved organic matter (1/day)

Ks3 = The degradation constant for suspended organic matter (1/day)

Kb3 = The degradation constant for deposited organic matter (1/day)

Biological Oxygen Demand, BOD

.d3 = Arrhenius temperature coefficient for dissolved organic matter

.s3 = Arrhenius temperature coefficient for suspended organic matter

.b3 = Arrhenius temperature coefficient for deposited organic matter

S1 = Resuspension rate for BODb (g/m2/day)

K5 = Deposition rate for BODs (m/day)

Ucrit = Critical velocity for deposition (m/s)

Vcrit = Critical BODb-concentration for resuspension ($g/m^2/day$)

10.4 Recommended Values

The values of the BOD decay rates depend on the nature of the organic matter. Organic matter can originate from household as well as industrial wastewater. The processes of wastewater treatment plants will also influence the degradability of the effluents. The most readily degradable components will decompose in the treatment plant depending on the processes in the treatment plant in question. The reported range of BOD decay rates is given by: 0.1 - 1.5 (1/day)

(Jørgensen, 1979). The variability of the degradability of industrial wastewater is extensive, ranging from relatively easily degradable waste from food industries to more persistent wastewater from e.g. pulp mills. As examples, the decay rate of BOD for wastewater from a sugar refinery was found to be 0.75 (1/day) (Water Quality Institute, 1984, see References) and for a sulphite pulp mill to be 0.25 (1/day) (Nyholm *et. al.*, 1991. Reported ranges of the temperature coefficient for degradation of organic matter in water are 1.02 - 1.09 (Jørgensen, 1979. A typical value would be 1.07. The suspended organic matter (BOD) in the environment at some distance from the outlet point consists of particles in the range 1 - 10 μm. The deposition rates for this range of particles of organic material lie between 0.07 - 0.7 (m/day). A typical value would be 0.2 (m/day) corresponding to a particle size of 5 μm. Typical values for the resuspension rate of BOD from the bed are difficult to specify. The rate has to be determined by calibrating the model to measured values or from experience of resuspension in similar situations.

10.5 Water Quality Modeling (Sag Curves Streeter and Phelps equation)

DO content is one of the most widely used indicators of overall ecological health of a body of water

- fish need 4 to 5 mg/L to survive
- under anaerobic conditions, undesirable (smelly) microbes can take over
- many factors affect the DO level

If a river is healthy before we begin discharging wastewater, a significant factor in its continued health or illness is the BOD added to it by wastewater

At the outfall, BOD of the river/wastewater mixture (L_0) is given by:

$$L_0 = \frac{Q_r L_r + Q_w L_w}{Q_r + Q_w} \quad \text{.............. 10.1}$$

Where

L_0 = Ultimate BOD at the point of waste discharge

Q_r = Flow in the river upstream of the discharge

L_r = Ultimate BOD of the river wate r

Q_w = Flow of wastewate r from the discharge

L_w = Ultimate BOD in the discharged wastewater

This BOD is comparable to what we have in our stoppered bottle at the beginning of our BOD test

As time passes (ie, the water moves downstream) the oxygen content of the river water is consumed in just the same way oxygen is consumed in the test Remember that BOD (L_t) in a test bottle at time t is given by:

$$L_t = L_0 e^{-kDt} \quad \text{.... 10.2}$$

This formula holds in the river too (k_D is the deoxygenation constant that we previously just called k; it can be adjusted for temperature using $k_T = k_{20q}{}^{T-20)}$ If we know an average velocity of flow, we can calculate the BOD for a given distance downstream, it is probably more interested in how much DO remains, which depends both on the rate of deoxygenation and on the rate of reoxygenation or reaeration.

- The rate of reaeration, r_R, is given by:
  $r_R = -k_R{}^*D$ with
  k_R = reaeration time constant
  D = DO deficit = DO_s-DO
- The reaeration time constant can be estimated from Table 10.2, or calculated by:
  $k_{R,20°C} = 3.9u^{1/2}/H^{3/2}$
  u = average stream velocity
  H = average stream depth

Table 10.2 Reaeration constants

Water body	**Ranges of k_R at 20°C, base e**
Small ponds and backwaters	0.1-0:23
Sluggish streams and large lakes	0.23-0.35
Large streams of low velocity	0.35-0.46
Large streams of normal velocity	0.46-0.69
Swift streams	0.69-1.15
Rapids and waterfalls	Greater than 1.15

Source : Peavy, Rowe and Tchobanoglous, 1985

To start with, the waste has some oxygen deficit which causes an initial DO deficit in the stream. The Reaeration constant s which are used in this chapter are given in table 10.1 Water can only hold so much oxygen (DO_{sat}), depending on the the water temperature. Calculate the initial dissolved oxygen (DO_0) using the same formula we used for L_0 above. Subtracting that from the initial DO_{sat}:

$$D_0 = DO_{sat} - DO_0 \qquad \text{......... 10.3}$$

The DO at any point downstream depends on these competing processes: rate of deficit increase = rate of deoxygenation - rate of reaeration This gives us a differential equation with the solution:

$$D = \frac{k_D L_0}{k_R - k_D}\left(e^{-k_D t} - e^{k_R t}\right) + D_0 e^{-k_R t} \qquad \text{......... 10.4}$$

This is the Streeter-Phelps oxygen-sag curve formula. Note that for a constant stream cross-section, t=x/u (with u=stream velocity); therefore:

If we want to plot DO versus distance downstream we need to subtract D from D_s at each point

To start with, DO is being depleted faster than it can be replenished. As long as this occurs, the DO of the stream will continue to drop. Since the BOD is decreasing as time goes on, at some point, the rate of deoxygenation decreases to just the rate of reaearation. At this point (called the critical point) the DO reaches a minimum. Downstream of the critical point, reaeration occurs faster than deoxygenation, so the DO increases. Using calculus and the Streeter-Phelps equation we get:

$$t_c = \frac{1}{k_R - k_D} \ln\left(\frac{k_R}{k_D}\left[1 - \frac{D_0 (k_R - k_D)}{k_D L_0}\right]\right) \qquad \text{......... 10.5}$$

The first order decay rate for BOD can be estimated from degradation experiments. The standard measurement of BOD is the BOD5. Given the results of intermediate measurements (after 1, 2, 3 and 4 days) and a value that is believed to represent the ultimate BOD level, the decay rate can be extracted. The calculation is based on the equation:

BODT = BOD8 . (1 -e-KBOD. T)

BODT = BOD measured after T days

BOD8 = The ultimate BOD value

Linear regression on BOD8. BODT.

It provides the KBOD value. It can very often be assumed that the BOD7 ~ BOD8 . The validity of this assumption can be checked by plotting the measured values as demonstrated in the sketch and/or by studying the regression coefficient (the higher the value the better the validity of the assumption).

10.6 Dissolved Oxygen, DO

Photosynthesis takes place during the day time only. The actual day length automatically calculated by MIKE 11 depends on the time of the year and the latitude. Concurrent with the oxygen producing photosynthesis is the oxygen consuming respiration by plants, bacteria and animals. Whereas the oxygen production only occurs in daytime, the respiration processes continue throughout the day and night. The respiration is temperature dependent described by the Arrhenius expression and an Arrhenius temperature coefficient .2. In addition to the temperature coefficient, the respiration rate at 20.C has to be given.

The WQ model at levels 3, 4 and 6 includes the oxygen consumption due to oxidation of ammonia to nitrate (the nitrification process). The nitrification is described as a first order process.

The decay of ammonia gives rise to an oxygen demand of two moles of oxygen per mole of nitrogen oxidised. Multiplying this with the moleweights of oxygen (O_2) and nitrogen (N) gives the 'yield' factor describing the amount of oxygen used at nitrification. The reaeration process expresses the reaeration related to the saturation concentration of oxygen. At concentrations lower than the saturation level, oxygen is transferred from the air to the water phase at a specific rate. If the concentration becomes higher than the saturation level, oxygen will similarly be transferred to the air. The reaeration rate (the transfer rate) is determined from water depth, water velocity and river slope.

Due to the large variation in velocity, depth and slope of rivers and streams, six different expressions for the re aeration constant have been implemented. The user must choose from these. The first expression is recommended for streams, the second for ordinary rivers, the third for rivers with high flow velocities and the fourth, fifth and sixth are expressions to be specified by the user. The model calculates the oxygen saturation concentration given the relevant temperature. The equations describing the reaeration coefficient and the oxygen saturation concentration are given under Processes.

The User specified values specific for the oxygen processes are summarized below:

All model levels

Pmax = Rate of oxygen production by photosynthesis ($gO2/m^2/day$).

R20 = Respiration rate at 20°C ($gO2/m^2/day$).

2 = Arrhenius temperature coefficient for respiration.

B1 = Sediment oxygen demand; the total oxygen demand for the model levels 2 and 4 and only the basic/natural oxygen demand ($gO2/m^2/day$) for the model levels 5 and 6.

.1 = Arrhenius temperature coefficient for oxygen demand.

Model levels 3, 4 and 6 only

Y1 = Factor describing the amount of oxygen consumed by nitrification (gO2/gNH3-N).

10.7 Recommended Values

The biological processes of photosynthesis and respiration show a yearly variation and, additionally, a diurnal variation of the photosynthetic oxygen production. The model determines the diurnal variation. The userspecified values required are the daily maximum of photosynthesis and the constant respiration (constant over the day). Typical values of oxygen production in the growth season will be 1.75 - 7.0 $gO2/m^2/day$ corresponding to a primary production of 0.5 - 2.0 $gC/m^2/day$. The respiration rate will typically be around 1.0 - 5.0 $gO2/ m^2/day$. A typical value for the temperature coefficient of respiration will be 1.08 (Jørgensen, 1979).

The natural sediment oxygen demand, e.g. the oxygen consumption from organic matter not originating from pollution sources, is typically in the range 0.2 - 1.0 $gO2/m^2/day$. These values are valid for 'sandy bottom' (Jørgensen, 1979). The same reference states values for 'estuary mud' and 'aged sewage sludge' in the range 1 - 2 $gO2/m^2/day$, but these types of sediment are definitely affected by pollution sources. A value of 0.5 $gO2/m^2/day$ is recommended for the natural sediment oxygen demand and a value of 1.5 $gO2/m^2/day$ for the total oxygen demand in the lower model levels. A temperature coefficient of 1.07 is typical. The yield factor for nitrification, i.e. the stoichiometric conditions and the molar weights determine the amount of oxygen consumed by nitrification. Two moles of oxygen are consumed per mole of ammonia oxidised. Multiplying with the molar weight of oxygen (32) and dividing by the molar weight of nitrogen (14) gives a yield factor of 4.57 (gO2/gNH3-N).

To specify the parameters for oxygen production, respiration rate and sediment oxygen demand, in units of gO2/m2/day, it is necessary to consider.

10.8 Definitions

The following section provides a very brief introduction to this topic, which is frequently encountered as part of addressing specific impacts within estuaries. The section is included so that these aspects are not forgotten but the reader is referred to the literature cited for more extensive coverage.

Water quality modelling solves the equation of advection-diffusion for a predefined computational grid on a wide range of model substances. Water quality models aim to provide the user with information regarding the "condition" of a water body. Typically, water quality models are not hydrodynamic models and require tidal information to drive the model.

The impact of a discharge on any water body is dependent on a discharge quantity and prevailing physical and chemical conditions. Typically, spatial and temporal conditions for a water ecosystem are highly variable. This is due to tidal, wind, currents, bathymetry, fluvial flow and so on. To asses the impact of a discharge it is necessary to predict the duration over which the pollutants may act and the area likely to be effected.

10.9 Defining the Model

It is essential to define the major issues and variables under consideration at the outset in order to select an appropriate model.

- *Model Duration :* Temporal extent of the discharge. It defines the simulation period which the model simulates processes.
- *Model Domain :* Spatial extent of the model determined from knowledge of the location and temporal effect of the discharge.

- *Model Dimensions :* Requires the knowledge of the hydrology of the area and behaviour of the pollutants.

Which then need to be mapped onto what the different types of model can deliver:

- *1 Dimensional Model :* Single scale, e.g. length down an estuary. Typically 1D modelling is undertaken for river networks and estuaries, which are well mixed where longitudinal variation is important
- *2 Dimensional Model :* Two scales, e.g. length and depth of an estuary. Typically, 2D water quality modelling is undertaken for water bodies, which are well mixed but have a significant width.
- *3 Dimensional Model :* Three scales, e.g. length, width and depth. If width and depth variation are important a 3D dynamic model is required.
- *Box Models :* These are simple representations of complex systems that focus on fluxes between boxes and transformations within boxes that are considered relatively homogeneous. In water quality modelling box models can be used in a variety of ways. For example, the simplest box model is the tidal prism model. However, a collection of simple box models can be linked together horizontally and vertically to represent the major features of an estuarine system. They are effective water quality management tools that can incorporate information from measured data or higher order two and three-dimensional models results (Ibrahim, 1996).
- *Inverse models :* This type of model is able to interpolate objectively between scattered data by making use of a system of governing equations (for example conservation equations for mass and momentum). They require no boundary data, thus they are only useful when only scattered observations rather than boundary data are available. An inverse model needs good data for assimilation (Lam *et al.* 1984; Copeland, 1994; Copeland & Bayne, 1999).

10.10 Modeled Substances

There is a wide range of substances that can be included within a water quality model. These include:

- Conservative substances (salinity, chloride, traces);
- Decayable substances;
- Suspended sediment;
- Temperature;
- Nutrients (ammonia, nitrate, phosphate, silicate);
- Organic Matter (nitrogen, carbon, phosphorus, silicon);
- Oxygen;
- BOD and COD (biological and chemical oxygen demand);
- Algae;
- Bacteria;
- Heavy Metals.

Water quality models allow the user to specify a wide range of physical, biological, ecological and chemical processes. These processes include;

- Sedimentation and resuspension;
- Reaeration of oxygen
- Algae growth and mortality
- Mineralisation of organic substances;
- Denitrification;
- Adsorption of heavy metals.

10.11 Model Calibration / Validation

Calibration is the process by which the model is adjusted to reproduce the characteristics of the study area for a given set of conditions (STOWA/RIZA, 1999) Water quality calibration consists of concentrations of the modelled substances at points throughout the model area over the period of interest. Seasonal variation may be important for some parameters such as nutrients and chlorophyll. In addition, it is important that all inputs to the model area from e.g. outfalls or rivers are accurately specified.

Model validation demonstrates the accuracy of the model output with a separate, independent dataset. Calibration and validation of a water quality model should reflect the timescales of the parameters in question, tidal and possibly seasonal variability. The modeling of biological oxygen demand is an interrelated part of the dissolved oxygen (DO) modeling. In this chapter described about the BOD and DO modeling of fresh water ecosystem.

References

Copeland GJM, 1994, An inverse method of kinematic flow modelling based on measured currents, *Proceedings of Institution of Civil Engineers, Water, Maritime & Energy*, 106(3), 249-258.

Copeland GJM, Bayne GLS, 1999, Tidal flow modelling using a direct minimisation method, *Coastal Engineering*, 34(1-2), 129-161.

Ibrahim ZZ, 1996, A spring-neap flushing box model, In: Pattiaratchi C (Ed.), Mixing in Estuaries and Coastal Seas, American Geophysical Union, Washington DC, pp. 278-290.

J~rgensen, S.E., 1976. A eutrophication model for a lake. *Ecol.Modelling*, 2: 147-165.

Lam DCL, Murthy CR, Simpson RB, 1984, Effluent transport and diffusion models for the coastal zone, Springer-Verlag, New York.

Nyholm, N., 1978. A simulation model for phytoplankton growth and nutrient recycling in eutrophic shallow lakes. *Ecol. Modelling*, 4: 279-310.

STOWA/RIZA, 1999, Smooth modelling in water management: Good modelling practice handbook, Dutch Department of Public Works, Institute of Inland Water Management and Waste Water Treatment, Den Haag, Netherlands, *Report No: STOWA report* 99-05, 1-167.

Chapter 11

Salinity and Temperature Modeling

11.1 Introduction

Salinity is the saltiness or dissolved salt content of a body of water. The technical term for saltiness in the ocean is halinity, from the fact that halides, chloride specifically are the most abundant anions in the mix of dissolved elements. In oceanography, it has been traditional to express halinity not as percent, but as parts per thousand (ppt), which is approximately grams of salt per liter of solution. In other disciplines chemical analyses of solutions, and thus salinity is frequently reported in mg/L or ppm (parts per million). Prior to 1978, salinity or halinity was expressed as ‰ usually based on the electrical conductivity ratio of the sample to "Copenhagen water", an artificial sea water manufactured to serve as a world "standard"(Lewis, E.L 1980). In 1978, oceanographers redefined salinity in Practical Salinity Units (psu): the conductivity ratio of a sea water sample to a standard KCl solution (UNESCO,1981a) (UNESCO, 1981b). Ratios have no units, so it is not the case that 35 psu exactly equals 35 grams of salt per litre of solution(UNESCO,1985).

These seemingly esoteric approaches to measuring and reporting salt concentrations may appear to obscure their practical use; but it must be remembered that salinity is the sum weight of many different elements within a given volume of water. It has always been the case that to get a precise salinity as a concentration and convert this to an amount of substance required knowing much more about the sample and the measurement than just the weight of the solids upon evaporation. For example, volume is influenced by water temperature; and the composition of the salts is not a constant. Saline waters from inland seas can have a composition that differs from that of the ocean. For the latter reason, these waters are termed saline as differentiated from ocean waters, where the term haline applies.

Marine waters are those of the ocean, another term for which is euhaline seas. The salinity of euhaline seas is 30 to 35 ‰. Brackish seas or waters have salinity in the range of 0.5 to 29‰ and metahaline seas from 36 to 40‰. These waters are all regarded as thalassic because their salinity is derived from the ocean and defined as homoiohaline if salinity does not vary much over time, modified from Por (1972) (Por, F.D 1972), follows the "Venice system" (1959).

In contrast to homoiohaline environments are certain poikilohaline environments in which the salinity variation is biologically significant (Dahl, E. 1956). Poikilohaline waters may range anywhere from 0.5‰ to greater than 300‰. The important characteristic is that these waters tend to vary in salinity over some biologically meaningful range seasonally or on some other roughly comparable time scale. Put simply, these are bodies of water with quite variable salinity.

11.2 Environmental Considerations

Salinity is an ecological factor of considerable importance, influencing the types of organisms that live in a body of water. As well, salinity influences the kinds of plants that will grow either in a water body, or on land fed by a water or by a groundwater. A plant adapted to saline conditions is called a halophyte. Organisms that can live in very salty conditions are classified as extremophiles, halophiles specifically. An organism that can withstand a wide range of salinities is euryhaline. Salt is difficult to remove from water, and salt content is an important factor in water use such as potability.

11.3 Temperature

The temperature of an ideal monatomic gas is a measure related to the average kinetic energy of its atoms as they move.

Temperature is a physical property of a system that underlies the common notions of hot and cold; something that is hotter generally has the greater temperature. Temperature is one of the principal parameters of thermodynamics. The temperature of a system is defined as simply the average energy of microscopic motions of a single particle in the system per degree of freedom. For a solid, these microscopic motions are principally the vibrations of the constituent atoms about their sites in the solid. For an ideal monatomic gas, the microscopic motions are the translational motions of the constituent gas particles. For multiatomic gas vibrational and rotational motion should be included too.

Temperature is measured with thermometers that may be calibrated to a variety of temperature scales. In most of the world, the Celsius scale is used for most temperature measuring purposes. The entire scientific world measures temperature using the Celsius scale and thermodynamic temperature using the Kelvin scale, which is just the Celsius scale shifted downwards (Lewis, E.L 1980)=-273.15°C, or absolute zero. Many engineering fields in the U.S., especially high-tech ones, also use the Kelvin and Celsius scales. The bulk of the U.S. however, relies upon the Fahrenheit scale. Other engineering fields in the U.S. also rely upon the Rankine scale when working in thermodynamic related disciplines such as combustion.

11.4 Overview

Intuitively, temperature is a measure of how hot or cold something is. On the molecular level, temperature is the result of the motion of particles which make up a substance. Temperature increases as the energy of this motion increases. The motion may be the translational motion of the particle, or the internal energy of the particle due to molecular vibration or the excitation of an electron energy level. Although very specialized laboratory equipment is required to directly detect the translational thermal motions, thermal collisions by atoms or molecules with small particles suspended in a fluid produces Brownian motion that can be seen with an ordinary microscope. The thermal motions of atoms are very fast and temperatures close to absolute zero are required to directly observe them. For instance, when scientists at the NIST achieved a record-setting cold temperature of 700 nK (1 nK = 10^{-9} K) in 1994, they used optical lattice laser equipment to adiabatically cool caesium atoms.

They then turned off the entrapment lasers and directly measured atom velocities of 7 mm per second in order to calculate their temperature. Molecules, such as O_2, have more degrees of freedom than single atoms: they can have rotational and vibrational motions as well as translational motion. An increase in temperature will cause the average translational energy to increase. It will also cause the energy associated with vibrational and rotational modes to increase. Thus a diatomic gas, with extra degrees of freedom like rotation and vibration, will require a higher energy input to change the temperature by a certain amount, i.e. it will have a higher heat capacity than a monatomic gas.

The process of cooling involves removing energy from a system. When there is no more energy able to be removed, the system is said to be at absolute zero, which is the point on the thermodynamic temperature scale where all kinetic motion in the particles comprising matter ceases and they are at complete rest in the "classic" sense. By definition, absolute zero is a temperature of precisely 0 kelvins (-273.15 °C or -459.67 °F).

The formal properties of temperature follow from its mathematical and are studied in thermodynamics and statistical mechanics.Contrary to other thermodynamic quantities such as entropy and heat, whose microscopic definitions are valid even far away from thermodynamic equilibrium, temperature being an average energy per particle can only be defined at thermodynamic equilibrium, or at least local thermodynamic equilibrium.

As a system receives heat, its temperature rises; similarly, a loss of heat from the system tends to decrease its temperature.

When two systems are at the same temperature, no heat transfer occurs between them. When a temperature difference does exist, heat will tend to move from the higher-temperature system to the lower-temperature system, until they are at thermal equilibrium. This heat transfer may occur via conduction, convection or radiation or combinations of them.

Temperature is also related to the amount of internal energy and enthalpy of a system: the higher the temperature of a system, the higher its internal energy and enthalpy are.

Temperature is an intensive property of a system, meaning that it does not depend on the system size, the amount or type of material in the system, the same as for the pressure and density. By contrast, mass, volume, and entropy are extensive properties, and depend on the amount of material in the system.

11.5 The Role of Temperature in Nature

Temperature plays an important role in almost all fields of science, including physics, chemistry, and biology.Many physical properties of materials including the phase (solid, liquid, gaseous or plasma), density, solubility, vapor pressure, and electrical conductivity depend on the temperature. Temperature also plays an important role in determining the rate and extent to which chemical reactions occur. This is one reason why the human body has several elaborate mechanisms for maintaining the temperature at 37 °C, since temperatures only a few degrees higher can result in harmful reactions with serious consequences. Temperature also controls the type and quantity of thermal radiation emitted from a surface. One application of this effect is the incandescent light bulb, in which a tungsten filament is electrically heated to a temperature at which significant quantities of visible light are emitted.

11.6 Temperature Measurement

Temperature measurement using modern scientific thermometers and temperature scales goes back at least as far as the early 18th century, when Gabriel Fahrenheit adapted a thermometer (switching to mercury) and a scale both developed by Ole Christensen Rømer. Fahrenheit's scale is still in use, alongside the Celsius scale and the Kelvin scale.

11.7 Units of Temperature

The basic unit of temperature (symbol: T) in the International System of Units (SI) is the kelvin (Symbol: K). The Kelvin and Celsius scales are, by international agreement, defined by two points: absolute zero, and the triple point of Vienna Standard Mean Ocean Water (water specially prepared with a specified blend of hydrogen and oxygen isotopes). Absolute zero is defined as being precisely 0 K and -273.15 °C. Absolute zero is where all kinetic motion in the particles comprising matter ceases and they are at complete rest in the "classic" (non-quantum mechanical) sense. At absolute zero, matter contains no thermal energy. Also, the triple point of water is defined as being precisely 273.16 K and 0.01 °C. This definition does three things: i) it fixes the magnitude of the kelvin unit as being precisely 1 part in 273.16 parts the difference between absolute zero and the triple point of water; ii) it establishes that one kelvin has precisely the same magnitude as a one degree increment on the Celsius scale; and iii) it establishes the difference between the two scales' null points as being precisely 273.15 kelvins (0 K = -273.15 °C and 273.16 K = 0.01 °C). Formulas for converting from these defining units of temperature to other scales can be found.

11.8 Temperature Conversion

In the field of plasma physics, because of the high temperatures encountered and the electromagnetic nature of the phenomena involved, it is customary to express temperature in electronvolts (eV) or kiloelectronvolts (keV), *where* 1 eV = 11,604 K. In the study of QCD matter one routinely meets temperatures of the order of a few hundred MeV, equivalent to about 10^{12} K.

For everyday applications, it's very often convenient to use the Celsius scale, in which 0 °C corresponds to the temperature at which water freezes and 100 °C corresponds to the boiling point of water at sea level. In this scale a temperature difference of 1 degree is the same as a 1 K temperature difference, so the scale is essentially the same as the Kelvin scale, but offset by the temperature at which water freezes (273.15 K).

11.9 Modeling - Salinity

Wastewater discharges of brines are associated with oil and gas drilling and production and desalination facility discharge. In the oil and gas industry they are commonly referred to as "*produced waters*". Because these flows tend to have greater density than the surrounding ambient waters, they are negatively buoyant and will sink towards the bottom. After bottom boundary interaction density current mixing is likely to occur using CORMIX model.

Care should be exercised when simulating these flows within software. Although the system does recognize negatively buoyant flow classes the system assumes flat bottom topography. In the far-field as density currents, these flows will have a strong tendency to travel down slope, following local bathymetric contours. (Fig. 11.2)

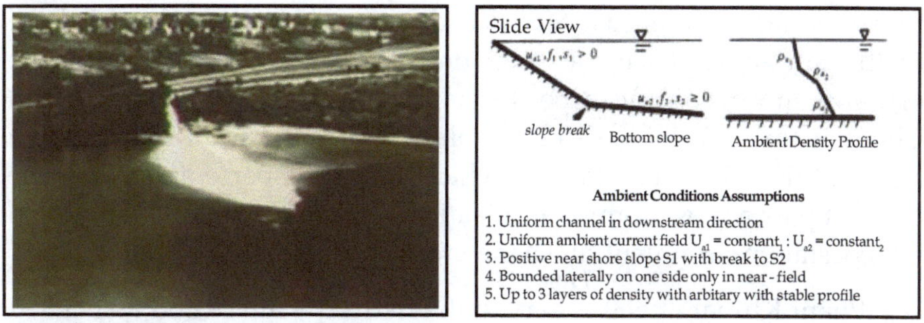

Fig. 11.2 Surface Brine Discharge on the Freshwater Lake Produces a Plunging Density Current.

11.10 Case study - Ennore Coast

The coast is one of the most important boundaries in the world. It separates the marine part of the world from the continents. It separates the salt-water communities from fresh water communities, it separates land life from water borne life but amphibians living both environments. Some of the marine organisms move towards estuary for reproduction with less water salinity. The coastline as such is not a stable. It fluctuates in place over considerable distances due to erosion and sedimentation, but also due to the relative change of sea level with respect to land. The coastal zone is thus a very mobile part of the earth from a geographical point of view. Due to its specific geophysical properties, the coast is also a very important area for the development of flora and fauna. There is hardly any other part of the world that houses such a variety of species, both in flora and fauna.

The abundance of opportunities along the coast has also attracted a lot of human activities. The coastal zone has become an indispensable resource for mankind. It provides excellent opportunities for fisheries, housing, recreation, transport, defense purpose, industrial activities and many others. The influx of man with their social and economic activities, their needs for safe housing, food, drinking water, and with the tremendous quantities of waste they are producing poses at the same time a serious threat to the unique features of the coastal zone. It is therefore necessary to study this valuable region so as to prevent that competing activities are gradually causing depletion of the resources in the coastline.

Chennai is situated near the coast of Bay of Bengal. It is potential for commercial and industrial activities and has witnessed rapid growth of industries during past few decades. Oceans cover two-thirds of earth's surface and have a water depth of over 4000m. The effluent contains more salinity to be discharged in to coastal area of north Chennai for a Reverse Osmosis plant. Indrani Gupta *et. al.* (2002) reported that the tidal currents which advert disperse and dilute the effluent play a most dominant role in governing the pollutant levels in the coastal environment at Thane creek, Mumbai.

Phani Kumar (1993) observed the hydrodynamics and associated pollutant transport in Thane creek. In his model, the bottom friction is a constant and model results were compared only with a limited data set. Vaze et. al. (2002) presented extensive studies on selection of location of intake and outfall for a power plant. In their selection, the intake is located at Ennore port basin and outfall near the creek mouth. It restored the imbalance in the tidal influx and efflux. Hegde A.V. (2001) reported the impact assessment of XCF outfall discharge at Mangalore coast using CORMIX model. It is assessed 3 marine species for Lethal Concentration for 50 organisms (LC_{50}) Criteria Maximum Concentration (CMC) and Criteria Continuous Concentration (CCC). One species is unsafe during and after slack tide and two species are safe after slack tide. Purendare U.V. et.al.(2001) studied near field dilution performance of multiport diffusers using CORMIX model. From the experiments logical trends have been observed in respect of dilution, spread etc. both qualitatively and quantitatively. Yan *et. al.* (1993) reported various similarity criteria developed from the momentum equation of the variable density flow and *experimentally examined* regarding the effect of model designs. Another study for a Petroleum Corporation is situated near north Chennai. It is proposed to install a Reverse Osmosis Plant to cater for their existing refinery including the expansion requirements.

The proposal includes a seawater intake and discharge system in the vicinity of North Chennai (Ennure) from / to the Bay of Bengal. The discharge brine water from the Reverse Osmosis plant is to be discharged at suitable location in the sea such that saline recirculation would be avoided at the intake location. There is no rise in temperature of effluent at the point of discharge.

The following environmental conditions revealed the suitability of salinity dispersion in the study area.

Tides and tidal currents

The tides are semi-diurnal with low amplitude. The maximum spring tidal range is 0.6 m while minimum neap tidal range is 0.2 m. The tidal currents and direction were observed in 35m depth of water (NIOT, 2002) at north Chennai that were used for these studies. The minimum current speed ranged between 0.6 cm/s and 1.39 cm/s, the maximum current speed ranged between 52.54 cm/s and 63cm/s. The current direction changes from south to north during February, and from February to September, the coastal currents are northerly. It is seen that northerly currents are stronger (about 0.6m/s) compared to southerly currents (about 0.1m/s). The current direction generally has the tendency towards north and south, parallel to the shoreline.

Salinity

The average seawater salinity is 32ppt and the effluent containing 63ppt and 80ppt when the recovery is 35% and 50% respectively from the reverse osmosis plant.

Temperature

The surface temperature distribution in the area exhibits the presence of warm water towards the coast and relatively cold water 28.5 °C offshore. The near bottom temperature varies from 29.3 °C at the coast to 28.5 °C offshore. The maximum seawater temperature variation of 1.1 °C in the nearshore water can be attributed to the release of warm water from the nearby thermal power station.

11.11 Modeling -Temperature

Most aquatic organisms are cold-blooded meaning they are unable to internally regulate their core body temperature. Therefore, temperature exerts a major influence on the biological activity and growth of aquatic organisms. To a point, the higher the water temperature, the greater the biological activity. Fish, insects, zooplankton, phytoplankton, and other aquatic species all have preferred temperature ranges. As temperatures get too far above or below this preferred range, the number of individuals of the species decreases until finally there are few, or none. For example, we would generally not expect to find a thriving trout fishery in ponds or shallow lakes because the water is too warm throughout the ice-free season.

The modeling of temperature should be done for safety of human life on the earth.

11.12 A Case study - Kalpakkam Coast

Kalpakkam is located on the east coast of India, about 70 km south of Chennai in Tamil Nadu. Mahabalipuram, the famous tourist place is about 5 km north of Kalpakkam. Edaiyur creek mouth is situated on north of the proposed power station. The historical Buckingham canal connects the Edaiyur backwaters on the north with the Sadras creek system on the south, near Sadhurangpattinam.

The sea bed slopes are steep (1:20) in the nearshore region and flat(1:100) in the region beyond. During the period of observations, tides are semi-diurnal with low amplitude. The maximum spring tidal range during the period is 1.22 m while the minimum neap tidal range is 0.25 m. The tidal currents are quite weak and do not exhibit any correlation with the tide or tidal phase. The tidal currents are of the order of 3 cm/s. The currents are more influenced by the bay circulations generated by southwesterly and northeast monsoons. The coastal currents are southerly during October to January. The current direction changes from south to north during February and from February to September, the coastal currents are northerly.

The combined action of warm water discharge from a power station and littoral drift resulted in the formation of sand-spit and about 2000 m long channel parallel to the shore. The warm water flowing in the channel gets discharged in the sea at the tip of the sand-spit. The length of the sand-spit varies corresponding to the change in the magnitude and direction of the littoral drift. As such, the point of the warm water meeting the sea varies during different seasons.

The warm water discharge not only punctures the sand spit at several locations at a time but also cuts the sand-spit at its root and warm water flows in the direction of the intake-well of power station. The distribution of temperature is strongly influenced by the seasonally reversing coastal currents at this coastal area. The thermal plume moves to the north hugging the coast when the currents are northerly during the southwest monsoon period (June-September) and moves to the south when the currents are southerly during the northeast monsoon period (October-February). There are also situations, particularly during the transition from northeast to southwest monsoon, when the thermal plume showed a movement towards the power station intake. The effect of the hot water discharge is felt up to 2 m below water surface. The seawater at the bottom did not show any effect of the discharge of warm water. Due to small magnitude of tidal currents, the dispersion characteristics at the site are quite low causing the influence of the warm water over a considerably large area.

For these studies, mathematical models OUTRAY, MIKE21-HD and MIKE21-AD, CORMIX are used for simulation of nearshore wave field, assessment of littoral drift at the site, coastal hydrodynamics and dispersion of warm water in the nearshore region respectively. The OUTRAY model is developed by HR, Wallingford, UK. The LITPACK, MIKE21-HD and MIKE21-AD are developed by Danish Hydraulics Institute, Denmark.

$$S_t + P_x + Q_y = 0 \qquad \text{......... 11.1}$$

and CORMIX are widely used commercial software. Brief description of these models is given below.

For determining the nearshore wave climate at the proposed site at Kalpakkam, by transforming the ship based deep water wave data, mathematical model OUTRAY was used. This model is based on wave ray technique. It takes into account refraction and shoaling of waves, which are important in the transformation of waves from offshore to inshore. The model takes into account directional waves by applying cos-squared directional spreading function to unidirectional spectrum.

The wave height and period are input to the model by specifying the spectral distribution of wave energy at the offshore boundary. For this purpose Pierson-Moskowitz uni directional spectrum is considered at the offshore boundary. The model converts this unidirectional spectrum to multidirectional spectrum by applying cos-squared distribution as follows –

$$S(f,q) = (2/p)\ S(f)\ \cos^2 (q-q_0) \qquad \text{......... 11.2}$$

where,

$S(f,q)$ = multidirectional energy spectrum

$S(f)$ = unidirectional spectrum

f = frequency

q = wave direction

q_0 = predominant wave direction

In the OUTRAY model, wave rays are tracked backwards from the inshore point of interest (Refraction Point) to the offshore boundary and a transfer function for wave conditions at the Refraction Point with respect to the offshore boundary is generated. Using the transfer function and the incident wave conditions at the offshore boundary, the wave climate at the inshore location is obtained.

The hydrodynamic model is used for simulation of water levels and flows in coastal areas. It simulates unsteady two-dimensional flows in coastal area and is based on the following non-linear vertically integrated 2-D equations of conservation of mass and momentum.

$$P_t + \left(\frac{P^2}{h}\right)_x + \left(\frac{PQ}{h}\right)_y + ghS_x + \frac{gP\sqrt{P^2+Q^2}}{C^2h^2} - \frac{1}{\rho_w}\left[(h\tau_{xx})_x + (h\tau_{xy})_y\right] - \Omega Q - fVV_y + \frac{h}{\rho_w}(p_a)_x = 0 \qquad .....11.3$$

where, S=surface elevation (m), P, Q = flux densities in x, y directions (m^2/s), h=water depth (m), C=Chezy resistance ($m^{1/2}/s$), V, V_x, V_y = wind speed and components in x, y directions (m/s), f=wind friction factor, Ω=coriolis parameter, p_a=atmospheric pressure, τ_{xx}, τ_{yy}, τ_{xy}=components of effective shear stress(kg/m^2), x,y = space coordinates (m), t= time (s). These equations are numerically solved by

$$Q_t + \left(\frac{Q^2}{h}\right)_y + \left(\frac{PQ}{h}\right)_x + ghS_y + \frac{gQ\sqrt{P^2+Q^2}}{C^2h^2} - \frac{1}{\rho_w}\left[(h\tau_{yy})_y + (h\tau_{xy})_x\right] - \Omega P - fVV_x + \frac{h}{\rho_w}(p_a)_y = 0 \qquad ..... 11.4$$

Alternate Direction Implicit (ADI) finite difference technique leading to formation of tri-diagonal matrices, which can be solved efficiently.

The Advection-Dispersion model simulates dispersion of warm water in coastal waters. It is based on the following non-linear vertically integrated 2-D equation of conservation of heat which takes into account advection, dispersion, heat loss to atmosphere, warm water source and sink.

$$(hT)_t + (uhT)_x + (vhT)_y = (hD_x\ T_x\ )_x + (hD_y\ T_y\ )_y - F\ h\ T + S \qquad ..... 11.5$$

where, T = rise in temperature (^{0}C), u,v =horizontal velocity components in x, y directions (m/s), D_x , D_y =dispersion coefficients in x, y directions (m^2/s), F = heat decay coefficient(s^{-1}) , S = source/ sink discharge ($m^3/s/m^2$). This equation is numerically solved by an explicit finite difference scheme.

In the case of artificially warmed aquatic area, the loss of heat will increase due to long wave radiation, evaporation and convection. This increase in heat loss is included as a decay term FhT, where F is a heat decay coefficient. The following expression is used for computing the decay coefficient.

$$F = 0.2388/(\rho.C_p.H)(4.6 - 0.09(T_r + T) + 4.06.W) \quad \exp(0.033(T_r + T)) \quad ..... 11.6$$

where, ρ = density of water(kg/m^3), C_p = specific heat, H = water depth (m), T_r = reference temperature(^{0}C), T = excess temperature (^{0}C), W = wind speed (m/s).

2D models are depth averaged in which the variation of parameters over the vertical is not simulated. However, since it is observed from field measurements that the effect of hot water discharge is felt up to about 2 m from the surface of water, the dispersion of warm water is restricted to depth of 2 m. Rainfall and cyclonic conditions are also considered in this modeling. The model simulates the far field advection-dispersion of the warm water. In case of water, conduction is insignificant while advection/convection is important and is considered in the model.

This chapter is discussed about meaning, definitions, naturally occurring, pollutant source and their effects on ecosystem of salinity and temperature parameters. Modeling of salinity and temperature in coastal area are discussed.

The salinity and temperature modeling are studied with the help of case studies. It is well discussed about the model and results of the modeling studies.

References

Dahl, E. 1956. Ecological salinity boundaries in poikilohaline waters. *Oikos*, 7(I): 1-21.

Lewis, E.L. 1980. The Practical Salinity Scale 1978 and its antecedents. *IEEE J. Ocean. Eng.*, OE-5(1): 3-8.

Mantyla, A.W. 1987. Standard Seawater Comparisons updated. *J. Phys. Ocean.*, 17: 543-548

MIKE21 and MIKE11,LITPACK,DHI, Denmark, release 2003

Por, F. D. 1972. Hydrobiological notes on the high-salinity waters of the Sinai Peninsula. *Mar. Biol.*, 14(2): 111–119.

UNESCO 1981a. The Practical Salinity Scale 1978 and the International Equation of State of Seawater 1980. *Tech. Pap. Mar. Sci.*, 36: 25 pp.

UNESCO 1981b. Background papers and supporting data on the Practical Salinity Scale 1978. *Tech. Pap. Mar. Sci.*, 37: 144 pp.

UNESCO 1985. The International System of Units (SI) in Oceanography. *Tech. Pap. Mar. Sci.*, 45: 124 pp.

Venice system (1959). The final resolution of the symposium on the classification of brackish waters. *Archo Oceanogr. Limnol.*, 11 (suppl): 243–248.

□□□

Chapter 12

Nitrogen and Phosphorus Modeling

12.1 Introduction-Nitrogen

Nitrogen is a chemical element which has the symbol N and atomic number 7. Elemental nitrogen is a colourless, odourless, tasteless and mostly inert diatomic gas at standard conditions, constituting 78.1% by volume of Earth's atmosphere. Nitrogen is a constituent element of all living tissues and amino acids. Many industrial important compounds, such as ammonia, nitric acid, and cyanides are made up of nitrogen.

12.1.1 History

Nitrogen (Latin *nitrogenium,* where *nitrum* (from Greek *nitron*) means "native soda"and *genes* means "forming") is formally considered to have been discovered by Daniel Rutherford in 1772, who called it *noxious air* or *fixed air.* That there is a fraction of air that did not support combustion is well known to the late 18th century.

Argon was discovered when it is noticed that nitrogen from air is not identical to nitrogen from chemical reactions. Compounds of nitrogen were known in the Middle Ages. The alchemists knew nitric acid as *aqua fortis* (strong water). The mixture of nitric and hydrochloric acids is known as *aqua regia* (royal water), celebrated for its ability to dissolve gold (the *king* of metals). The earliest industrial and agricultural applications of nitrogen compounds involved uses in the form of saltpeter (sodium- or potassium nitrate), notably in gunpowder, and much later, as fertilizer, and later still, as a chemical feedstock.

Solid nitrogen ice in a small plastic beaker with melting liquid flowing off. The nitrogen has been frozen by immersion in liquid helium.

Nitrogen gas is acquired for industrial purposes by the fractional distillation of liquid air, or by mechanical means using gaseous air. Commercial nitrogen is often a byproduct of air-processing for industrial concentration of oxygen for steel making and other purposes.

Nitrogen gas has a wide variety of applications, including serving as an inert replacement for air where oxidation is undesirable;

- To preserve the freshness of packaged or bulk foods (by delaying rancidity and other forms of oxidative damage)
- In ordinary incandescent light bulbs as an inexpensive alternative to argon
- On top of liquid explosives for safety
- The production of electronic parts such as transistors, diodes, and integrated circuits
- Dried and pressurized, as a dielectric gas for high voltage equipment
- The manufacturing of stainless steel
- Use in military aircraft fuel systems to reduce fire hazard,
- Filling automotive and aircraft tires[4] due to its inertness and lack of moisture or oxidative qualities, as opposed to air, though this is not necessary for consumer automobiles. (Cartalk,2007) (Straight Dope, 2007).

Nitrogen molecules are less likely to escape from the inside of a tire compared with the traditional air mixture used. Air consists mostly of nitrogen and oxygen. Nitrogen molecules have a larger effective diameter than oxygen molecules and therefore diffuse through porous substances more slowly (Van Amerongen 1946).

A further example of its versatility is its use as a preferred alternative to carbon dioxide to pressurize kegs of some beers, particularly thicker stouts and Scottish and English ales, due to the smaller bubbles it produces, which make the dispensed beer smoother and headier. A modern application of a pressure sensitive nitrogen capsule known commonly as a "widget" now allows nitrogen charged beers to be packaged in cans and bottles.

Molecular nitrogen, a diatomic gas, is apt to dimerize into a linear four nitrogen long polymer. This is an important phenomenon for understanding high voltage nitrogen dielectric switches because the process of polymerization can continue to lengthen the molecule to still longer lengths in the presence of an intense electric field. A nitrogen polymer fog is thereby created. The second virial coefficient of nitrogen also shows this effect as the compressibility of nitrogen gas is changed by the dimerization process at moderate and low temperatures.

Nitrogen tanks are also replacing carbon dioxide as the main power source for paintball markers. The downside is that nitrogen must be kept at higher pressure than CO_2, making N_2 tanks heavier and more expensive.

12.1.2 *Properties*

Nitrogen is a nonmetal, with an electronegativity of 3.0. It has five electrons in its outer shell and is therefore trivalent in most compounds. The triple bond in molecular nitrogen (N_2) is one of the strongest in nature. The resulting difficulty of converting (N_2) into other compounds, and the ease of converting nitrogen compounds into elemental N_2, have dominated the role of nitrogen in both nature

At atmospheric pressure molecular nitrogen condenses (liquifies) at 77 K (-195.8 °C) and freezes at 63 K (-210.0 °C) into the beta hexagonal close-packed crystal allotropic form. Below 35.4 K (-237.6 °C) nitrogen assumes the alpha cubic crystal allotropic form. Liquid nitrogen, a fluid resembling water, but with 80.8% of the density, is a common cryogen.

Unstable allotropes of nitrogen consisting of more than two nitrogen atoms have been produced in the laboratory, like N_3 and N_4. Under extremely high pressures and high temperatures (2000 K), as produced under diamond anvil conditions, nitrogen polymerizes into the single bonded diamond crystal structure, an allotrope nicknamed "nitrogen diamond." (physorg.com, 2004).

12.1.3 Occurrence

Nitrogen is the largest single component of the Earth's atmosphere (78.082% by volume of dry air, 75.3% by weight in dry air). ^{14}N is created as part of the fusion processes in stars, and is estimated to be the 7th most abundant chemical element in our universe.

Nitrogen is present in all living tissues as proteins, nucleic acids and other molecules. It is a large component of animal waste (for example, guano), usually in the form of urea, uric acid, and compounds of these nitrogenous products.

12.1.4 Isotopes

There are two stable isotopes of nitrogen: ^{14}N and ^{15}N. By far the most common is ^{14}N (99.634%), which is produced in the CNO cycle in stars and the remaining is ^{15}N. Of the ten isotopes produced synthetically, ^{13}N has a half life of ten minutes and the remaining isotopes have half lives on the order of seconds or less. Biologically-mediated reactions (e.g., assimilation, nitrification, and denitrification) strongly control nitrogen dynamics in the soil. These reactions typically result in ^{15}N enrichment of the substrate and depletion of the product. 0.73% of the molecular nitrogen in Earth's atmosphere is comprised of the isotopologue $^{14}N^{15}N$ and almost all the rest is $^{14}N_2$.

12.1.5 Electromagnetic spectrum

Molecular nitrogen ($^{14}N_2$) is largely transparent to infrared and visible radiation because it is a homonuclear molecule and thus has no dipole moment to couple to electromagnetic radiation at these wavelengths. Significant absorption occurs at extreme ultraviolet wavelengths, beginning around 100 nanometers. This is associated with electronic transitions in the molecule to states in which charge is not distributed evenly between nitrogen atoms. Nitrogen absorption leads to significant absorption of ultraviolet radiation in the earth's upper atmosphere as well as in the atmospheres of other planetary bodies. For similar reasons, pure molecular nitrogen lasers typically emit light in the ultraviolet range. Nitrogen also makes a contribution to visible air glow from the earth's upper atmosphere, through electron impact excitation followed by emission. This visible blue air glow typically results not from molecular nitrogen, but rather from free nitrogen atoms combining with oxygen to form nitric oxide.

12.1.6 *Liquid nitrogen*

Liquid nitrogen is a cryogenic liquid. At atmospheric pressure, it boils at -196 °C. When insulated in proper containers such as dewar flasks, it can be transported without much evaporative losses.

Like dry ice, the main use of liquid nitrogen is as a refrigerant. Among other things, it is used in the the cryopreservation of blood, reproductive cells, and other biological samples and materials. It is also used in cold traps for certain laboratory equipment. It has also been used to cool central processing units and other devices in computers which are overclocked, and which produce more heat than during normal operation.

12.1.7 *Biological role*

Nitrogen is an essential part of amino acids and nucleic acids, both of which are essential to all life on earth. Molecular nitrogen in the atmosphere cannot be used directly by either plants or animals, and needs to be converted to other compounds, or "fixed," in order to be used by life. Precipitation often contains substantial quantities of ammonium and nitrate, both thought to be a result of nitrogen fixation by lightning and other atmospheric electric phenomena. However, because ammonium is preferentially retained by the forest canopy relative to atmospheric nitrate, most of the fixed nitrogen that reaches the soil surface under tree canopy is in the form of nitrate. Soil nitrate is preferentially assimilated by tree roots relative to soil ammonium.

Specific bacteria (Rhizobium *trifolium*) possess nitrogenase enzymes which can fix atmospheric nitrogen into a form which is chemically useful to higher organisms. This process requires a large amount of energy and anoxic conditions. Such bacteria may be free in the soil (azotobacter) but normally exist in a symbiotic relationship in the root nodules of leguminous plants. Nitrogen fixating bacteria can be symbiotic with a number of unrelated plant species. Common examples are legumes, alders, lichens, casuarina, myrica, liverwort, and gunnera.

As part of the symbiotic relationship, the plant subsequently converts the ammonium ion to nitrogen oxides and amino acids to form proteins and other biologically useful molecules, such as alkaloids. In return for the usable nitrogen, the plant secretes sugars to the symbiotic bacteria.

Some plants are able to assimilate nitrogen directly in the form of nitrates which may be present in soil from natural mineral deposits, artificial fertilizers, animal waste, or organic decay. Nitrates absorbed in this fashion are converted to nitrites by the enzyme nitrate reductase, and then converted to ammonia by another enzyme called nitrite reductase.

Nitrogen compounds are basic building blocks in animal biology. Animals use nitrogen-containing amino acids from plant sources, as starting materials for all nitrogen-compound animal biochemistry, including the manufacture of proteins and nucleic acids. Some plant-feeding insects are so dependent on nitrogen in their diet, that varying the amount of nitrogen fertilizer applied to a plant can affect the birth rate of the insects feeding on it. (Jahn *et al,* 2005)

Soluble nitrate is an important limiting factor in the growth of certain bacteria in ocean waters. In many places in the world, artificial fertilizers applied to crop-lands to increase yields result in run-off delivery of soluble nitrogen to oceans at river mouths. This process can result in eutrophication of the water, as nitrogen-driven bacterial growth depletes water oxygen to the point that all higher organisms die. Well-known "dead zone" areas and the Black Sea are due to this important polluting process.

Many saltwater fish manufacture large amounts of trimethylamine oxide to protect them from the high osmotic effects of their environment. In animals, the free radical molecule nitric oxide (NO), which is derived from an amino acid, serves as an important regulatory molecule for circulation.

Animal metabolism of NO results in production of nitrite. Animal metabolism of nitrogen in proteins generally results in excretion of urea, while animal metabolism of nucleic acids results in excretion of urea and uric acid. The characteristic odour of animal flesh decay is caused by nitrogen containing long chain amines, such as putrescine and cadaverine.

Nitrogen is a constituent of molecules in every major drug class in pharmacology and medicine. Nitrous oxide (N_2O) is discovered early in the 19th century to be a partial anesthetic, though it is not used as a surgical anesthetic until later. Called "laughing gas", it is found capable of inducing a state of social disinhibition resembling drunkenness. Other notable nitrogen-containing drugs are drugs derived from plant alkaloids, such as morphine. Nitrogen containing drugs include all of the major classes of antibiotics, and organic nitrate drugs like nitroglycerin and nitroprusside which regulate blood pressure and heart action by mimicking the action of nitric oxide.

12.1.8 Impacts of Nitrogen

The most important effect of deposited nitrogen on the flora and fauna of fresh waters is caused through acidification and mobilisation of aluminium. Although it is difficult to disentangle the contributions of deposited sulphur and nitrogen to acidification, there is no doubt that nitrogen has contributed significantly to changes in the biota in many parts of the India.

Increasing the nitrate nutrient content of fresh water usually has relatively little effect on the flora and fauna. This is because nitrogen, although low in most fresh waters, is not usually the limiting nutrient. Phosphorus is most commonly limiting and it is the most frequent cause of eutrophication of fresh water. Increasing the phosphorus supply tends to encourage the growth of cyanobacteria (which can fix their own nitrogen). Such "algal blooms" can be toxic to fauna, farm animals and humans.

However, there is still concern about nitrogen eutrophication of fresh water because research in the Netherlands has shown that shallow, soft water lakes and pools often are nitrogen limited. When the nitrogen concentration is increased, rare submerged plants are threatened and the flora is dominated by rushes. We need a much better picture of where nitrogen limitation occurs in India waters and of the potential threats to macrophytes.

High nitrate concentrations in water are associated with intensive agriculture and to a lesser extent to sewage discharges. In the early 1980s it was realized that the nitrate content of drinking water was high in certain lowland agricultural areas of the India. You can see the change in nitrate concentration in the Ganges between 1930 and 1997 at the Environment Agency.

At the time it is thought that high concentrations might increase the risk of certain cancers but that is no longer considered to be the case. It was also known that very high nitrates may cause a condition called methaemoglobinaemia in babies. That triggered a great deal of preventative action, as you will see later, but the condition is very rare and the last recorded case in the India was in 1992. The actual risk to human health posed by current levels of nitrate appears to be very small.

Nitrogen also dissolves in the bloodstream and body fats, and rapid can lead to a potentially fatal condition called decompression sickness, when nitrogen bubbles form in the bloodstream, nerves, joints, and other sensitive or vital areas.

Direct skin contact with liquid nitrogen causes severe frostbite within seconds, though not instantly on contact, depending on form of liquid nitrogen and surface area of the nitrogen-soaked materials.

12.2 Introduction- Phosphorus

Phosphorus, Greek: *phôs* meaning "light", and *phoros* meaning "bearer"), is the chemical element that has the symbol P and atomic number 15. A multivalent nonmetal of the nitrogen group, phosphorus is commonly found in inorganic phosphate rocks.

Due to its high reactivity, phosphorus is never found as a free element in nature. One form of phosphorus (white phosphorus) emits a faint glow upon exposure to oxygen (hence its Greek derivation and the Latin light-bearer meaning the planet Venus as dawn-bearer or "Morning Star"). Phosphorus is a component of DNA and RNA and essential element for all living cells. The most important commercial use of phosphorus-based chemicals is the production of fertilizers. Phosphorus compounds are also widely used in explosives, nerve agents, friction matches, fireworks, pesticides, toothpaste, and detergents.

12.2.1 Characteristics and allotropes

Elemental phosphorus as prepared artificially exists in several allotropes, most commonly white, red and black.White phosphorus

(P_4) exists as individual molecules made up of only four atoms in a tetrahedral arrangement, resulting in very high ring strain and instability. It contains 6 single bonds.

White phosphorus is a yellow, waxy transparent solid. For this reason it is also called yellow phosphorus. It glows greenish in the dark, is highly flammable and pyrophoric upon contact with air as well as toxic. The odor of combustion of this form has a characteristic garlic smell, and samples are commonly coated with white "(di)phosphorus pentoxide", which actually consists of P_4O_{10} tetrahedra with oxygen inserted between the phosphorus atoms and at their vertices. White phosphorus is insoluble in water but soluble in carbon disulfide.

Red phosphorus may be formed by heating white phosphorus to 250°C (482°F) or by exposing white phosphorus to sunlight. Phosphorus after this treatment exists as an amorphous network of atoms which reduces strain and gives greater stability, further heating results in the red phosphorus becoming crystaline. Red phosphorus does not catch fire in air at temperatures below 240°C whereas white phosphorus ignites at about 40°C.

Black phosphorus has an orthorhombic structure (C_{mca}) and is the least reactive allotrope, it consists of many six membered rings which are interlinked. Each atom is bonded to three other atoms.(Howstuffworks,2006) A recent synthesis of black phosphorus using metal salts as catylsts has been reported.

12.2.2 *Glow*

The glow from phosphorus is the attraction of its discovery around 1669, but the mechanism for that glow was not fully described until 1974. (Straight Dope, 2007). It is known from early times that the glow would persist for a time in a stoppered jar but then cease. Robert Boyle in the 1680s ascribed it to "debilitation" of the air. In fact it is oxygen being consumed. By the 18th century it was known that in pure oxygen phosphorus does not glow at all, (Van Amerongen 1946) there is only a range of partial pressure where it does, too high or too low and the reaction stops. Heat can be applied to drive the reaction at higher pressures (Jahn *et al*, 2005).

In 1974 the glow is explained by R. J. van Zee and A. U. Khan. (Straight Dope,2007). A reaction with oxygen takes place at the surface of the solid (or liquid) phosphorus, forming the short-lived molecules HPO and P_2O_2 that both emit visible light. The reaction is slow and only very little of the intermediates is required to produce the luminescence, hence the extended time the glow continues in a stoppered jar.

Although the term phosphorescence is derived from phosphorus, the reaction is properly called luminescence.

12.2.3 Applications

Concentrated phosphoric acids, which can consist of 70% to 75% P_2O_5 are very important to agriculture and farm production in the form of fertilizers. Global demand for fertilizers led to large increases in phosphate (PO_4^{3-}) production in the second half of the 20th century. Other uses;

- Phosphates are utilized in the making of special glasses that are used for sodium lamps.
- Boneash, calcium phosphate, is used in the production of fine china.
- Sodium tripolyphosphate made from phosphoric acid is used in laundry detergents in several countries, and banned for this use in others.
- Phosphoric acid made from elemental phosphorus is used in food applications such as soda beverages. The acid is also a starting point to make food grade phosphates. These include mono-calcium phosphate which is employed in baking powder and sodium tripolyphosphate and other sodium phosphates. Among other uses these are used to improve the characteristics of processed meat and cheese. Others are used in toothpaste. Trisodium phosphate is used in cleaning agents to soften water and for preventing pipe/boiler tube corrosion.
- Phosphorus is widely used to make organophosphorus compounds, through the intermediates phosphorus chlorides and the two phosphorus sulfides: phosphorus pentasulfide, and phosphorus sesquisulfide. Organophosphorus compounds have many applications, including in plasticizers, flame retardants, pesticides, extraction agents, and water treatment.

- Phosphorus is also an important component in steel production, in the making of phosphor bronze, and in many other related products.
- White phosphorus is used in military applications as incendiary bombs, for smoke-screening as smoke pots and smoke bombs, and in tracer ammunition.
- Red phosphorus is essential for manufacturing matchbook strikers, flares, safety matches, pharmaceutical grade and street methamphetamine, and is used in cap gun caps.
- Phosphorus sesquisulfide is used in heads of strike-anywhere matches.
- In trace amounts, phosphorus is used as a dopant for N-type semiconductors.
- ^{32}P and ^{33}P are used as radioactive tracers in biochemical laboratories.

12.2.4 *Biological role*

Phosphorus is a key element in all known forms of life. Inorganic phosphorus in the form of the phosphate PO_4^{3-} plays a major role in biological molecules such as DNA and RNA where it forms part of the structural framework of these molecules. Living cells also utilize phosphate to transport cellular energy via Adenosine Tri Phosphate (ATP). Nearly every cellular process that uses energy gets it in the form of ATP. ATP is also important for phosphorylation, a key regulatory event in cells. Phospholipids are the main structural components of all cellular membranes. Calcium phosphate salts are used by animals to stiffen their bones. An average person contains a little less than 1 kg of phosphorus, about three quarters of which is present in bones and teeth in the form of apatite. A well-fed adult in the industrialized world consumes and excretes about 1-3 g of phosphorus per day in the form of phosphate. Phosphorus is an essential macromineral, which is studied extensively in soil conservation in order to understand plant uptake from soil systems.

In ecological terms, phosphorus is often a limiting nutrient in many environments, i.e. the availability of phosphorus governs the rate of growth of many organisms. In ecosystems an excess of phosphorus can be problematic, especially in aquatic systems.

12.2.5 History

Phosphorus (Greek *phosphoros* is the ancient name for the planet Venus, but in Greek mythology, Hesperus and Eosphorus could be confused with Phosphorus) is discovered by German alchemist Hennig Brand in 1669 through a preparation from urine, which naturally contains considerable quantities of dissolved phosphates from normal metabolism. Working in Hamburg, Brand attempted to distill some salts by evaporating urine, and in the process produced a white material that glowed in the dark and burned brilliantly. Since that time, phosphorescence has been used to describe substances that shine in the dark without burning.

Phosphorus is first made commercially, for the match industry, in the 19th century, by distilling off phosphorus vapour from precipitated phosphates heated in a retort. The precipitated phosphates are made from ground up bones that had been de-greased and treated with strong acids. This process become obsolete in the late 1890s when the electric arc furnace was adapted to reduce phosphate rock.

Early matches used white phosphorus in their composition, which was dangerous due to its toxicity. Murders, suicides and accidental poisonings resulted from its use (Straight Dope,2007). In addition, exposure to the vapors give match workers a necrosis of the bones of the jaw, the infamous "phossy jaw." When a safe process for manufacturing red phosphorus is discovered, with its far lower flammability and toxicity, laws were enacted, under a Berne Convention, requiring its adoption as a safer alternative for match manufacture.

The electric furnace method allowed production to increase to the point phosphorus can be used in weapons of war. (Straight Dope,2007). In World War I it was used in incendiaries, smoke screens and tracer bullets. A special incendiary bullet was developed to shoot at hydrogen-filled Zeppelins over Britain. During World War II, Molotov cocktails of benzene and phosphorus are distributed in Britain to specially selected civilians within the British Resistance Operation, for defence; and phosphorus incendiary bombs are used in War on a large scale. Burning phosphorus is difficult to extinguish and if it splashes onto human skin it has horrific effects People covered in it have been known to commit suicide due to the torment.

Today phosphorus production is larger than ever, used as a precursor for various chemicals, (Fryzuk, M.D and Johnson, S.A 2000) in particular the herbicide glyphosate sold under the brand name Roundup. Production of white phosphorus takes place at large facilities and is transported heated in liquid form. Some major accidents have occurred during transportation, train derailments at Brownston, Nebraska and Miamisburg, Ohio lead to large fires. The worst accident in recent times though was an environmental one in 1968 when phosphorus spilt into the sea from a plant at Placentia Bay, Newfoundland.

12.2.6 Occurrence

Due to its reactivity to air and many other oxygen containing substances, phosphorus is not found free in nature but it is widely distributed in many different minerals. Phosphate rock, which is partially made of apatite, is an important commercial source of this element. Large deposits of apatite are located in China, Russia, Morocco, Florida, Idaho, Tennessee, Utah, and elsewhere. Albright and Wilson in the United Kingdom and their Niagara Falls plant, for instance, are using phosphate rock in the 1890s and 1900s from Connetable, Tennessee and Florida; however, by 1950 they are using phosphate rock mainly from Tennessee and North Africa. In the early 1990s Albright and Wilson's purified wet phosphoric acid business is being affected by phosphate rock sales by China and the entry of their long standing Moroccan phosphate suppliers into the purified wet phosphoric acid business.

Organic compounds of phosphorus form a wide class of materials, some of which are extremely toxic. Fluorophosphate esters are among the most potent neurotoxins known. A wide range of organophosphorus compounds are used for their toxicity to certain organisms as pesticides (herbicides, insecticides, fungicides etc) and weaponized as nerve agents. Most inorganic phosphates are relatively nontoxic and essential nutrients.

The allotrope, white phosphorus, should be kept under water at all times as it presents a significant fire hazard due to its extreme reactivity to atmospheric oxygen, and it should only be manipulated with forceps since contact with skin can cause severe burns. Chronic white phosphorus poisoning of unprotected workers leads to necrosis of the jaw called "phossy-jaw". Ingestion of white phosphorus may cause a medical condition known as "Smoking Stool Syndrome".

When the white form is exposed to sunlight or when it is heated in its own vapor to 250°C, it is transmuted to the red form, which does not phosphoresce in air. The red allotrope does not spontaneously ignite in air and is not as dangerous as the white form. Nevertheless, it should be handled with care because it does revert to white phosphorus in some temperature ranges and it also emits highly toxic fumes that consist of phosphorus oxides when it is heated.

Upon exposure to elemental phosphorus, in the past it is suggested to wash the affected area with 2% copper sulfate solution to form harmless compounds that can be washed away.

The manual suggests instead "a bicarbonate solution to neutralize phosphoric acid, which will then allow removal of visible WP. Particles often can be located by their emission of smoke when air strikes them, or by their phosphorescence in the dark. In dark surroundings, fragments are seen as luminescent spots." Then, "Promptly debride the burn if the patient's condition will permit removal of bits of WP which might be absorbed later and possibly produce systemic poisoning. Do not apply oily-based ointments until it is certain that all WP has been removed. Following complete removal of the particles, treat the lesions as thermal burns." As white phosphorus readily mixes with oils, any oily substances or ointments are disrecommended until the area is thoroughly cleaned and all white phosphorus removed.

Radioactive isotopes of phosphorus include

- ^{32}P; a beta-emitter (1.71 MeV) with a half-life of 14.3 days which is used routinely in life-science laboratories, primarily to produce radiolabeled DNA and RNA probes, *e.g.* for use in Northern blots or Southern blots. Because the high energy beta particles produced penetrate skin and corneas, and because any ^{32}P ingested, inhaled, or absorbed is readily incorporated into bone and nucleic acids, OSHA requires that a lab coat, disposable gloves, and safety glasses or goggles be worn when working with ^{32}P, and that working directly over an open container be avoided in order to protect the eyes. Monitoring personal, clothing, and surface contamination is also required. In addition, due to the high energy of the beta particles, shielding this radiation with the normally used dense materials (*e.g.* lead), gives rise to secondary emission of X-rays via a process known as Bremsstrahlung, meaning braking radiation. Therefore shielding must be accomplished with low density materials, *e.g.* Plexiglas, Lucite, plastic, wood, or water.[18]

- ^{33}P; a beta-emitter (0.25 MeV) with a half-life of 25.4 days. It is used in life-science laboratories in applications in which lower energy beta emissions are advantageous such as DNA sequencing.

12.3 Nitrogen and Phosphorus Modeling

An ocean model has been developed which, for prescribed physics, deals with interrelationships between chemical distributions, biogeochemical sinks and sources, chemical reactions at redox fronts, and transports across the air-sea and sediment-water interfaces. In its first application here, the model focuses on biogeochemical cycling of phosphorus, nitrogen, oxygen, and sulphur in an ocean forced by river input of nutrients. This is a natural starting point for a global climate model since ocean circulation and biology determine atmospheric CO_2 concentrations for a given inventory of inorganic C and oceanic production is controlled mainly by the availability of inorganic P and/ or N. A general approach is taken to look at oxic versus anoxic conditions. P versus N limitation of primary production, with or without inorganic removal of phosphate to the sediments. As demanded by this approach, the model is nonlinear and continuous in a vertical coordinate. To focus on the biogeochemical aspects, ocean physics are kept as simple as possible. Cold, oxygen-rich water sinks at high latitudes and is upwelled with a constant velocity. Turbulent mixing is parameterized with a constant, vertical diffusion coefficient. The biogeochemical processes considered are new production, burial, nitrogen fixation, phosphorite formation, and three types of organic decomposition: oxidation with O_2, denitrification, and sulphate reduction.

Organic matter is taken to consist of a high- and a low-reactive fraction. The chemical species considered explicitly are P_4^{3-}-P, NO_3^--N, O_2, NH_4^+-N and H_2S-S. Results indicate that a change from oxic to weakly anozic conditions at middepths in a P-limited ocean would lead to strong local denitrification and low nitrate concentrations throughout the water column. New production would also become dominated by nitrogen fixers. Geological evidence implies that anoxic conditions in the water column have been rare in the Phanerozoic ocean. Both phosphorite formation and denitrification can strongly constrain primary production and the development of anoxia. N limitation, i.e., negligable nitrogen fixation, practically precludes anoxia but is unlikely for very long time scales. For P limitation and no phosphorite formation the model indicates that the redox state of the ocean may be most sensitive to changes in ocean biology followed by changes in ocean circulation and mixing and finally by changes in ocean temperature. Nitrogen and phosphorus modeling, can be dealt with modeling chapter in more elaborately eutrophication.

References

A new molecule and a new signature - Chemistry - tetranitrogen. *Science News* (February 16, 2002). Retrieved on 2007-08-18.

Biology Safety - Cryogenic materials. The risks posed by them. University of Bath. Retrieved on 2007-01-03.

Diffusion, moisture and tyre expansion. Car Talk. Retrieved on 2006-07-22.

Fryzuk, M. D. and Johnson, S. A. 2000. "The continuing story of dinitrogen activation". *Coordination Chemistry Reviews* 200–202: 379. DOI:10.1016/S0010-8545(00)00264-2.

Fun with Cryogenics (part six) – video of the solid nitrogen experiment. You Tube (April 12, 2006).

G. J. Van Amerongen 1946. "The Permeability of Different Rubbers to Gases and Its Relation to Diffusivity and Solubility". *Journal of Applied Physics* 17 (11): 972–985. DOI:10.1063/1.1707667.

Is it better to fill your tires with nitrogen instead of air?. *The Straight Dope*. Retrieved on 2007-02-16.

Jahn, GC, LP Almazan, and J Pacia 2005. "Effect of nitrogen fertilizer on the intrinsic rate of increase of the rusty plum aphid, Hysteroneura setariae (Thomas) (Homoptera: Aphididae) on rice (*Oryza sativa* L.)". *Environmental Entomology* 34 (4): 938–943.

Polymeric nitrogen synthesised. *physorg.com* (August 5, 2004). Retrieved on 2007-08-18.

Why don't they use normal air in race car tires?. *Howstuffworks*. Retrieved on 2006-07-22.

❑❑❑

Chapter 13

Eutrophication Modeling

13.1 Introduction

Eutrophication, means an increase in chemical nutrients – typically compounds containing nitrogen and phosphorus in an ecosystem. It may occur on land or in water. The term is however often used to mean the resultant increase in the ecosystem's primary productivity (Fig. 13.1), in other words excessive plant growth and decay and even further impacts, including lack of oxygen and severe reductions in water quality and in fish and other animal populations.

Fig. 13.1. Eutrofied Lake and Coastal Ecosystem with Dense Bloom of Algae

Eutrophication is frequently a result of nutrient pollution such as the release of sewage effluent and run-off from lawn fertilizers into natural waters although it may also occur naturally in situations where nutrients accumulate or where they flow into systems on an ephemeral basis. Eutrophication generally promotes excessive plant growth and decay, favours certain weedy species over others, and is likely to cause severe reductions in water quality . In aquatic environments, enhanced growth of choking aquatic vegetation or phytoplankton disrupts normal functioning of the ecosystem, causing a variety of problems such as a lack of oxygen in the water, needed for fish and shellfish to survive.

The water then becomes cloudy, colored a shade of green, yellow, brown, or red. Human society is impacted as well: eutrophication decreases the resource value of rivers, lakes, and estuaries such that recreation, fishing, hunting, and aesthetic enjoyment are hindered. Health related problems can occur where eutrophic conditions interfere with drinking water treatment.(Bartram, J *et. al.* 1999)

Eutrophication is recognized as a pollution problem in indian lakes and reservoirs in the mid-20th century (Rodhe, W 1969). Since then, it has become more widespread. Surveys showed that 54% of lakes in Asia are eutrophic; in Europe, 53%; in North America, 48%; in South America, 41%; and in Africa, 28%. (ILEC, 1988-1993).

Eutrophication can be a natural process in lakes, as they fill in through geological time, though other lakes are known to demonstrate the reverse process, becoming less nutrient rich with time. Estuaries also tend to be naturally eutrophic because land-derived nutrients are concentrated where run-off enters the marine environment in a confined channel.

Human activities can accelerate the rate at which nutrients enter ecosystems (Fig. 13.2). Runoff from agriculture and development, pollution from septic systems and sewers, and other human-related activities increase the flux of both inorganic nutrients and organic substances into terrestrial, aquatic, and coastal marine ecosystems. Elevated atmospheric compounds of nitrogen can increase soil nitrogen availability.

Phosphorus is often regarded as the main culprit in cases of eutrophication in lakes subjected to point source pollution from sewage.

The concentration of algae and the trophic state of lakes correspond well to phosphorus levels in water. Studies conducted in the Experimental Lakes Area in Ontario have shown a relationship between the addition of phosphorus and the rate of eutrophication. Humankind has increased the rate of phosphorus cycling on Earth by four times, mainly due to agricultural fertilizer production and application. Between 1950 and 1995, 600,000,000 tonnes of phosphorus were applied to Earth's surface, primarily on croplands (Carpenter *et al.* 1998). Control of point sources of phosphorus have resulted in rapid control of eutrophication, mainly due to policy changes.

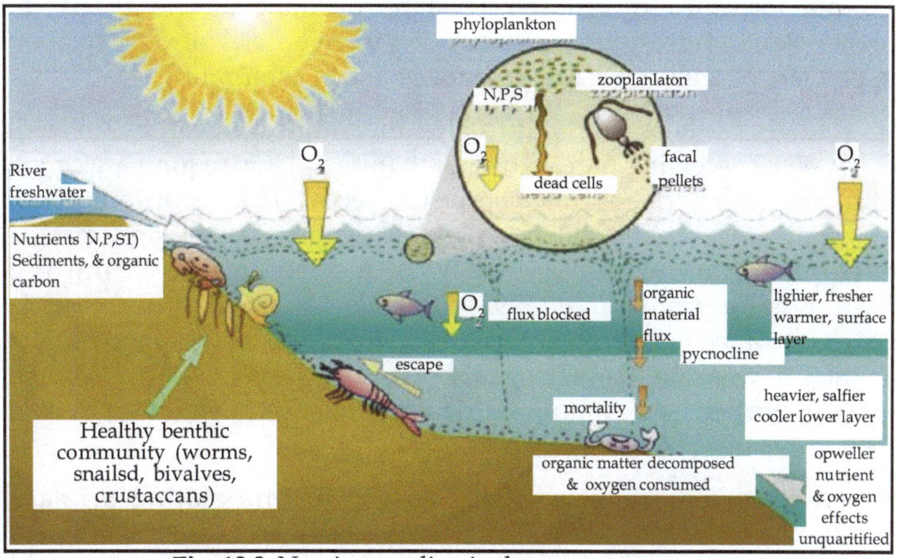

Fig. 13.2. Nutrient cycling in the water ecosystem

13.2 Terrestrial Ecosystems

Although traditionally thought of as enrichment of aquatic systems by addition of fertilizers into lakes, bays, or other semi enclosed waters (even slow-moving rivers), terrestrial ecosystems are subject to similarly adverse impacts (APIS 2005). Increased content of nitrates in soil frequently leads to undesirable changes in vegetation composition and many plant species are endangered as a result of eutrophication in terrestric ecosystems, e.g. majority of orchid species in Europe.

Ecosystems (like some meadows, forests and bogs that are characterized by low nutrient content and species rich, slowly growing vegetation adapted to lower nutrient levels) are overgrown by faster growing and more competitive species poor vegetation, like tall grasses, that can take advantage of unnaturally elevated nitrogen level and the area may be changed beyond recognition and vulnerable species may be lost. Eg. species-rich fens are overtaken by reed or reedgrass species, spectacular forest undergrowth affected by run-off from nearby fertilized field is turned into a thick nettle and bramble shrub.

Chemical forms of nitrogen are most often of concern with regard to eutrophication because plants have high nitrogen requirements so that additions of nitrogen compounds stimulate plant growth (primary production). Nitrogen is not readily available in soil because N_2, a gaseous form of nitrogen, is very stable and unavailable directly to higher plants. Terrestrial ecosystems rely on microbial nitrogen fixation to convert N_2 into other physical forms (such as nitrates).

However, there is a limit to how much nitrogen can be utilized. Ecosystems receiving more nitrogen than the plants require are called nitrogen saturated. Saturated terrestrial ecosystems contribute both inorganic and organic nitrogen to freshwater, coastal, and marine eutrophication, where nitrogen is also typically a limiting nutrient.(Hornunmg *et. al* 1995) However, because phosphorus is generally much less soluble than nitrogen, it is leached from the soil at a much slower rate than nitrogen. Consequently, phosphorus is much more important as a limiting nutrient in aquatic systems. (Smith et al 1999).

Many ecological effects can arise from stimulating primary production, but there are three particularly troubling ecological impacts: decreased biodiversity, changes in species composition and dominance, and toxicity effects.

- Increased biomass of phytoplankton
- Toxic or inedible phytoplankton species
- Increases in blooms of gelatinous zooplankton
- Decreased biomass of benthic and epiphytic algae
- Changes in macrophyte species composition and biomass
- Decreases in water transparency (increased turbidity)
- Color, smell, and water treatment problems
- Dissolved oxygen depletion

- Increased incidences of fish kills
- Loss of desirable fish species
- Reductions in harvestable fish and shellfish
- Decreases in perceived aesthetic value of the water body

13.3 Decreased Biodiversity

When an ecosystem experiences an increase in nutrients, primary producers reap the benefits first. In aquatic ecosystems, species such as algae experience a population increase. Algal blooms limit the sunlight available to bottom-dwelling organisms and cause wide swings in the amount of dissolved oxygen in the water. Oxygen is required by all respiring plants and animals and it is replenished in daylight by photosynthesizing plants and algae. Under eutrophic conditions, dissolved oxygen greatly increases during the day, but is greatly reduced after dark by the respiring algae and by microorganisms that feed on the increasing mass of dead algae. When dissolved oxygen levels decline to hypoxic levels, fish and other marine animals suffocate. As a result, creatures such as fish, shrimp, and especially immobile bottom dwellers die off. (Horrigan *et al* 2002) In extreme cases, anaerobic conditions ensue, promoting growth of bacteria such as *Clostridium botulinum* that produces toxins deadly to birds and mammals. Zones where this occurs are known as dead zones.

13.4 New Species Invasion

Eutrophication may cause competitive release by making abundant a normally limiting nutrient. This process causes shifts in the species composition of ecosystems. For instance, an increase in nitrogen might allow new, competitive species to invade and out-compete original inhabitant species. This has been shown to occur (Bertness et al 2002) in New England salt marshes.

13.5 Toxicity

Some algal blooms, otherwise called "nuisance algae" or "harmful algal blooms," are toxic to plants and animals. Toxic compounds they produce can make their way up the food chain, resulting in animal mortality.(Anderson D.M. 1994) Freshwater algal blooms can pose a threat to livestock. When the algae die or are eaten, neuro- and hepatotoxins are released which can kill animals and may pose a threat to humans. (Lawton L.A and G.A. Codd 1991) (Martin A and

G.D.Cooke 1994). An example of algal toxins working their way into humans is the case of shellfish poisoning.(Shumway S.E 1990) Biotoxins created during algal blooms are taken up by shellfish (mussels, oysters), leading to these human foods acquiring the toxicity and poisoning humans. Examples include paralytic, neurotoxic, and diarrhoetic shellfish poisoning. Other marine animals can be vectors for such toxins, as in the case of ciguatera, where it is typically a predator fish that accumulates the toxin and then poisons humans.

Nitrogen can also cause toxic effects directly. When this nutrient is leached into groundwater, drinking water can be affected because concentrations of nitrogen are not filtered out. Nitrate (NO_3) has been shown to be toxic to human babies. This is because bacteria can live in their digestive tract that convert nitrate to nitrite (NO_2). Nitrite reacts with hemoglobin to form methemoglobin, a form that does not carry oxygen. The baby essentially suffocates as its body receives insufficient oxygen. Characteristics of point and non-point sources of chemical inputs are as follows (Carpenter et al, 1998; Novonty and Olem 1994).

13.6 Point Sources

- Wastewater effluent (municipal and industrial)
- Runoff and leachate from waste disposal systems
- Runoff and infiltration from animal feedlots
- Runoff from mines, oil fields, unsewered industrial sites
- Overflows of combined storm and sanitary sewers
- Runoff from construction sites >20,000 m^2

13.7 Non-point Sources

- Runoff from agriculture/irrigation
- Runoff from pasture and range
- Urban runoff from unsewered areas
- Septic tank leachate
- Runoff from construction sites <20,000 m^2
- Runoff from abandoned mines
- Atmospheric deposition over a water surface
- Other land activities generating contaminants

13.8 Sources of High Nutrient Runoff

In order to gauge how to best prevent eutrophication from occurring, specific sources that contribute to nutrient loading must be identified. There are two common sources of nutrients and organic matter: point and non-point sources.

Point sources are directly attributable to one influence. In point sources the nutrient waste travels directly from source to water. For example, factories that have waste discharge pipes directly leading into a water body would be classified as a point source. Point sources are relatively easy to regulate.

Non-point source pollution is that which comes from ill-defined and diffuse sources. Non-point sources are difficult to regulate and usually vary spatially and temporally.

It has been shown that nitrogen transport is correlated with various indices of human activity in watersheds, (Cole J.J *et. al.* 1993 ; Howarth R.W *et. al.* 1996) including the amount of development. (Bertness et al 2002) Agriculture and development are activities that contribute most to nutrient loading. There are three reasons that nonpoint sources are especially troublesome (Smith *et al* 1999).

Nutrients from human activities tend to accumulate in soils and remain there for years. It has been shown (Sharpley *et. al.* 1996) that the amount of phosphorus lost to surface waters increases linearly with the amount of phosphorus in the soil. Thus much of the nutrient loading in soil eventually makes its way to water. Nitrogen, similarly, has a turnover time of decades or more.

Nutrients from human activities tend to travel from land to either surface or groundwater. Nitrogen in particular is removed through storm drains, sewage pipes, and other forms of surface runoff. Nutrient losses in runoff and leachate are often associated with agriculture. Modern agriculture often involves the application of nutrients onto fields in order to maximise production. However, farmers frequently apply more nutrients than are taken up by crops (Buol S.W 1995) or pastures. Regulations aimed at minimising nutrient exports from agriculture are typically far less stringent than those placed on sewage treatment plants (Carpenter *et al.*, 1998) and other point source polluters.

Nitrogen is released into the air because of ammonia volatilization and nitrous oxide production. The combustion of fossil fuels is a large human-initiated contributor to atmospheric nitrogen pollution. Atmospheric deposition (e.g., in the form of acid rain) can also effect nutrient concentration in water, (Paerl H.W1997) especially in highly industrialized regions.

Any factor that causes increased nutrient concentrations can potentially lead to eutrophication. In modeling eutrophication, the rate of water renewal plays a critical role; stagnant water is allowed to collect more nutrients than bodies with replenished water supplies. It has also been shown that the drying of wetlands causes an increase in nutrient concentration and subsequent eutrophication booms.(Mungall C and D.J. Mc Laren 1991)

Eutrophication poses a problem not only to ecosystems, but to humans as well. Reducing eutrophication should be a key concern when considering future policy, and a sustainable solution for everyone, including farmers and ranchers, seems feasible. While eutrophication does pose problems, humans should be aware that natural runoff (which causes algal blooms in the wild) is common in ecosystems and should thus not reverse nutrient concentrations beyond normal levels.

Cleanup measures have been mostly, but not completely, been successful. Finnish phosphorus removal measures started in the mid-1970s and have targeted rivers and lakes polluted by industrial and municipal discharges. These efforts have had a 90% removal efficiency.(Raike A *et. al* 2003) Still, some targeted point sources did not show a decrease in runoff despite reduction efforts.

Studies show that intercepting non-point pollution between the source and the water is a successful mean of prevention (Carpenter et al., 1998). Riparian buffer zones,an interface between a flowing body of water and land, have been created near waterways in an attempt to filter pollutants; sediment and nutrients are deposited here instead of in water. Creating buffer zones near farms and roads is another possible way to prevent nutrients from traveling too far. Still, studies have shown (Angold P.G 1997) that the effects of atmospheric nitrogen pollution can reach far past the buffer zone. This suggests that the most effective means of prevention is from the primary source.

Laws regulating the discharge and treatment of sewage have led to dramatic nutrient reductions to surrounding ecosystems, (Smith *et al.* 1999) but it is generally agreed that a policy regulating agricultural use of fertilizer and animal waste must be imposed. In Japan the amount of nitrogen produced by livestock is adequate to serve the fertilizer needs for the agriculture industry (Kumazawa K ,2002). Thus, it is not unreasonable to command livestock owners to clean up animal waste which when left stagnant will leach into groundwater.

13.9 Natural State of Algal Blooms

Although the intensity, frequency and extent of algal blooms has tended to increase in response to human activity and human-induced eutrophication, algal blooms are a naturally-occurring phenomenon. The rise and fall of algae populations, as with the population of other living things, is a feature of a healthy ecosystem. Rectification actions aimed at abating eutrophication and algal blooms are usually desirable, but the focus of intervention should not necessarily be aimed at eliminating blooms, but towards creating a sustainable balance that maintains or improves ecosystem health.

13.10 Eutrophication Modeling

The pollution of the Great Lakes from municipal and industrial sources was perceived to be a serious problem as early as the 1950's. The problem of eutrophication, including:

- Immediate reductions in the phosphorus content in detergents
- Implementation of programs aimed to reduce point source inputs (municipal and industrial) of phosphorus
- Development of programs for the control of phosphorus from agricultural sources
- Regulation of any new uses of phosphorus which could result in appreciable additions to the lakes.

The purpose of this chapter is to provide a history of the development and application of mathematical models of the eutrophication processes in a Lake and to comment on the potential use of eutrophication models in an ecosystem approach to address some other Lake water quality problems. However, we would be remiss if we did not include in this history the research and development efforts conducted in the other lakes of the Great Lakes system.

13.11 Model Evolution

The demonstration eutrophication model used seven vertically integrated segments and a steady-state circulation pattern to represent the physical and transport features of the western basin of Lake (Fig. 13.3 & 13.4). The estimates of the circulation pattern in the Lake are based on hydrodynamic model computations developed by Gedney (1971) and on observed current information reported by Herdendorf (1967). The kinetic framework employed for the eutrophication model (Fig. 13.3) incorporated eight state-variables (phytoplankton chlorophyll, herbivorous zooplankton carbon, carnivorous zooplankton carbon, organic and inorganic phosphorus, organic nitrogen, ammonia nitrogen and nitrate nitrogen) and followed the phytoplankton-zooplankton-nutrient model first structured by Di Toro *et al.* (1971). Using water quality data collected by the various investigators, water quality experts, the eutrophication model is calibrated with field data obtained from them. Initial calibration results were encouraging.

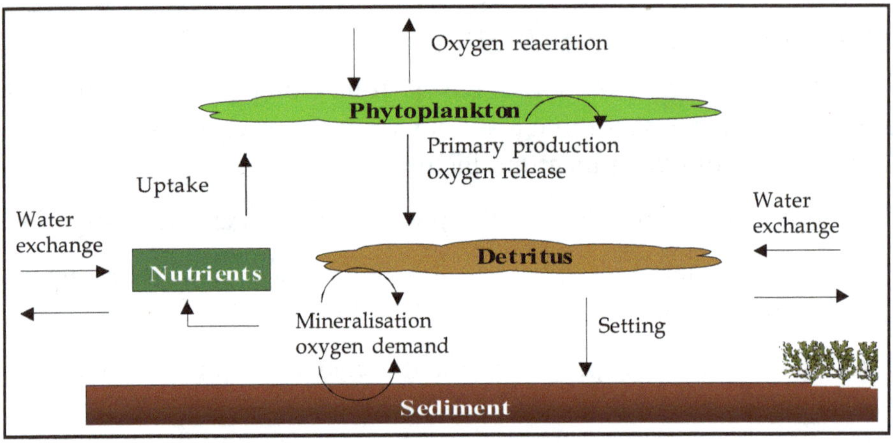

Fig. 13.3. Eutrophication Modeling (one layer)

The magnitudes and shapes of the calculated curves for phytoplankton biomass, as indicated by chlorophyll-a, inorganic phosphorus, ammonia nitrogen and nitrate nitrogen are in reasonable agreement with the observed data, although some systematic deviations were observed. The calibrated model is also used to hindcast lake water quality for the year 1930. Estimates of chlorophyll-a, used to validate the 1930 hindcast calculations, are based on observations of algal cells counts (Wright *et al.* 1955) and a relationship between chlorophyll-a and total algal cell counts. Model computations of chlorophyll-a for the 1930 hind cast compared reasonably to the observed data and provided an encouraging test of the Lake Erie western basin eutrophication model. Fig. 13.5 and Fig. 13.6 show the sub-module of nitrogen and phosphorus variables, which are used for the eutrophication modeling.

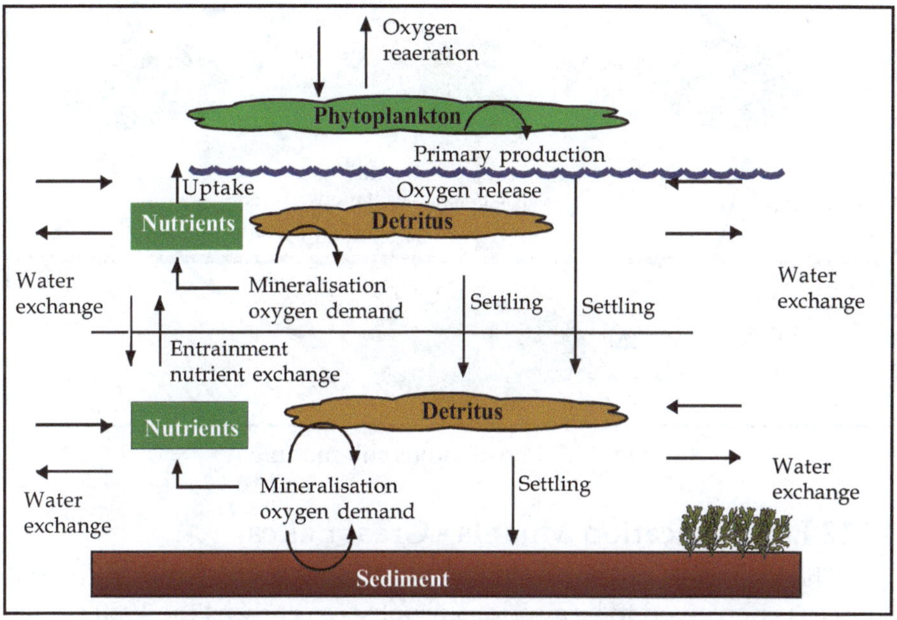

Fig. 13.4. Eutrophication Modelling (Two layers)

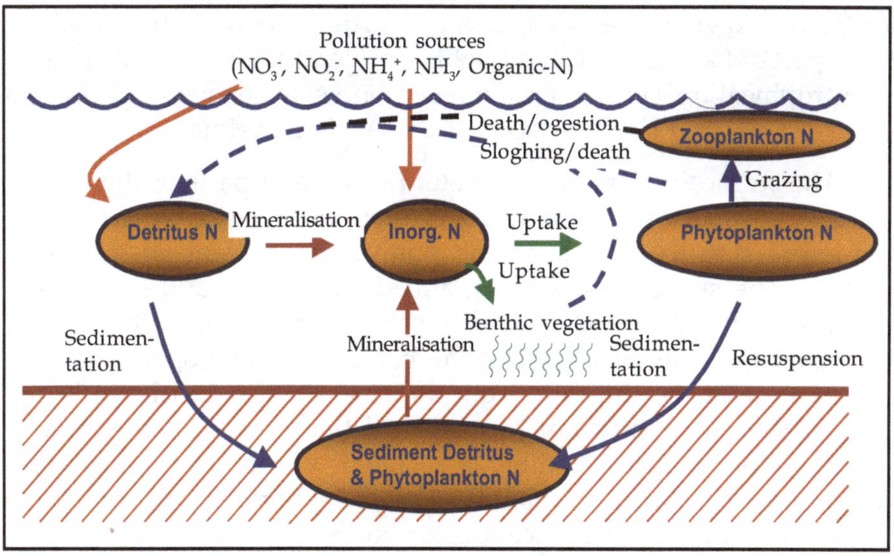

Fig. 13.5. Nitrogen sub-module

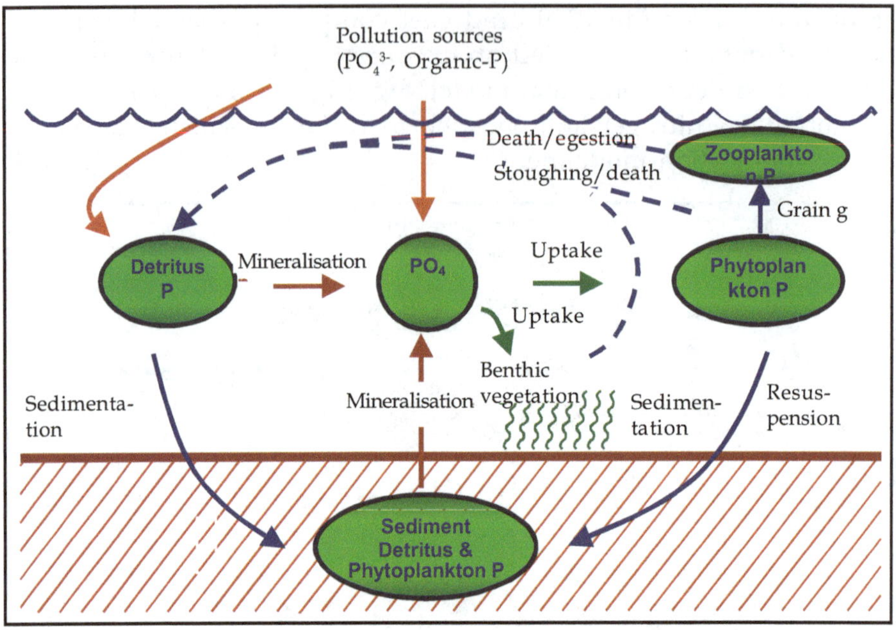

Fig. 13.6. Phosphorous sub-module

13.12 Eutrophication Models - Great Lakes

The pollution of the Great Lakes from municipal and industrial sources is perceived to be a serious problem as early as the 1950's. This perception is even greater for Lake Erie, which during the 1960's and 1970's is known as the "Dead Sea of North America." Historically, a large part of Lake Erie's water quality problems appeared to be related to eutrophication. During the summer months, windrows of *Cladophora glomerata* covered large portions of the lake's shoreline.

The surface waters of the western basin and parts of the central basin of Lake Erie are populated with *Aphanizomenon flosaquae*, giving the impression that someone had poured green paint on the water surface. The subsequent settling and decay of this and other algal biomass resulted in widespread hypoxia and anoxia in the bottom waters of the lake, further stressing commercial and sport fisheries, which are already suffering from the stresses of overfishing. While part of the eutrophication problems of the Great Lakes is due to increased phosphorus loadings associated with population growth, the post-World War II development of phosphorus-based detergents is also identified as a significant contributor to increased phosphorus loadings to the lakes. In its 1969 report to the United States and Canada, the International Joint Commission (IJC) concluded that phosphorus enrichment have accelerated the eutrophication of Lake Ontario and had resulted in a condition of advanced eutrophication in the western basin of Lake Erie. The IJC offered a number of recommendations to deal with the problem of eutrophication, including:

- Immediate reductions in the phosphorus content in detergents,
- Implementation of programs aimed to reduce point source inputs (municipal and industrial) of phosphorus,
- Development of programs for the control of phosphorus from agricultural sources,
- Regulation of any new uses of phosphorus which could result in appreciable additions to the lakes.

Subsequent to the 1969 IJC, the United States and Canada signed the Great Lakes Water Quality Agreement, which charted a course of action for the two governments in dealing with Great Lakes water quality problems. That Agreement, together with the 1972 Amendments of the Federal Water Pollution Control Act, which called for research and technical development with respect to Great Lakes water quality, including an analysis of present and projected future water quality under varying conditions of wastewater treatment and waste disposal, provided the impetus for water quality modeling of the Great Lakes.

The purpose of this chapter is to provide a history of the development and application of mathematical models of the eutrophication processes in Lake Erie and to comment on the potential use of eutrophication models in an ecosystem approach to address Lake Erie water quality problems. However, we would be remiss if we did not include in this history the research and development efforts conducted in the other lakes of the Great Lakes system.

One of the first eutrophication modeling studies of Lake Erie is reported by Di Toro *et al.* (1973). This work, which focused on Western Lake Erie, grew out work performed by Hydroscience (1973) for the Great Lake Basin Commission. The original Hydroscience study provided an assessment of the feasibility of applying a limnological systems analysis to the water resources problems of the Great Lakes. During the study, Hydroscience developed and calibrated a demonstration water quality model for western basin of Lake Erie.

The demonstration model included chlorides, coliform bacteria, eutrophication and a food chain model of cadmium.

The demonstration eutrophication model used seven vertically integrated segments and a steady-state circulation pattern to represent the physical and transport features of the western basin of Lake Erie. The estimates of the circulation pattern in western Lake Erie are based on hydrodynamic model computations developed by Gedney (1971) and on observed current information reported by Herdendorf (1967). The kinetic framework employed for the eutrophication model incorporated eight state-variables (Fig. 13.7) (phytoplankton chlorophyll, herbivorous zooplankton carbon, carnivorous zooplankton carbon, organic and inorganic phosphorus, organic nitrogen, ammonia nitrogen and nitrate nitrogen) and followed the phytoplankton-zooplankton-nutrient model first structured by Di Toro *et al.* (1971) for the Sacramento-San Joaquin Delta. Using water quality data collected by the Canadian Centre for Inland Waters (CCIW) and the USEPA, the eutrophication model is calibrated for the year 1970. Initial calibration results are encouraging.

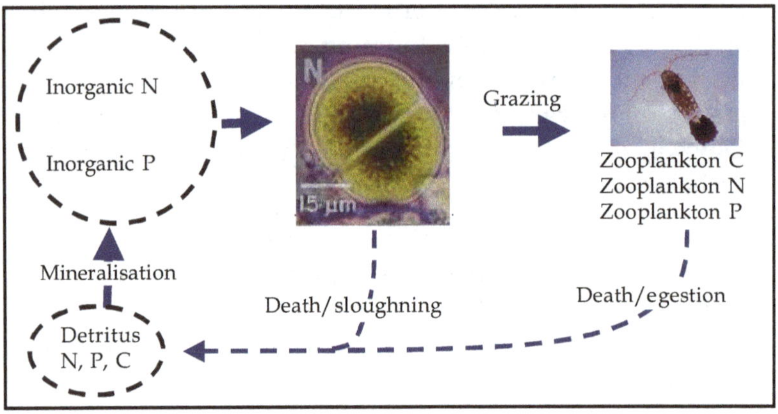

Fig. 13.7. Phytoplankton processes

The magnitudes and shapes of the calculated curves for phytoplankton biomass, as indicated by chlorophyll-a, inorganic phosphorus, ammonia nitrogen and nitrate nitrogen were in reasonable agreement with the observed data, although some systematic deviations were observed (Fig. 13.8). The calibrated model is also used to hindcast lake water quality for the year 1930. Estimates of chlorophyll-a, used to validate the 1930 hindcast calculations, are based on observations of algal cells counts collected by the U.S. Department of Interior, Fish and Wildlife Service for the period 1928-30 (Wright *et al.* 1955) and a relationship between chlorophyll-a and total algal cell counts based on EPA Western Lake Erie surveillance data for the period 1967-1968. Model computations of chlorophyll-a for the 1930 hindcast compared reasonably to the observed data and provided an encouraging test of the Lake Erie western basin eutrophication model.

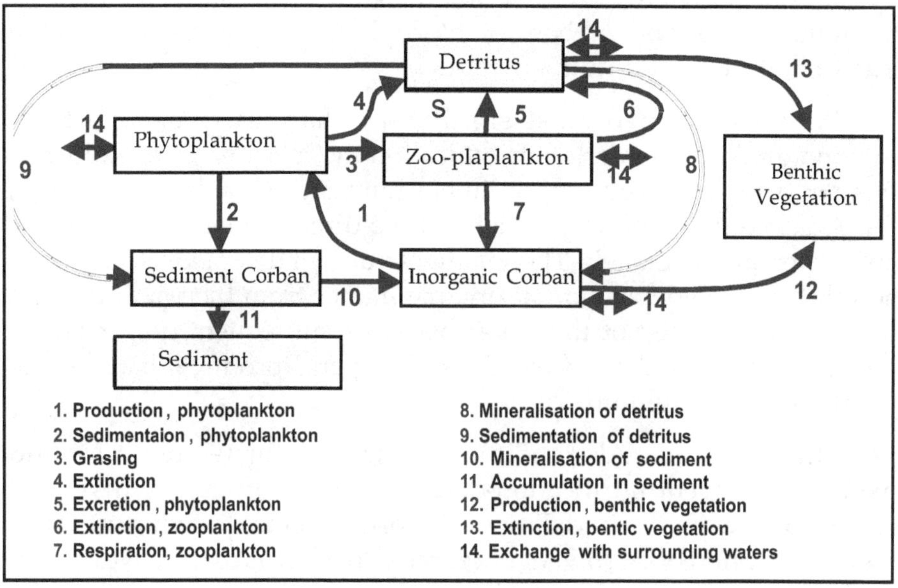

Fig. 13.8. Relationship of organic and inorganic materials in the ecosystems.

In many ways the current state-of-the-science in eutrophication modeling has not changed appreciably from the models of Di Toro, Thomann and Bierman. The basic linkages between nutrients, phytoplankton biomass, primary production, nutrient recycle and dissolved oxygen in today's eutrophication models are quite similar to the Great Lakes' models. Most modern eutrophication models still use the Monod theory for which algal growth rates depend upon external nutrient concentration rather than the formulations in which growth rate depends on the internal cellular nutrient concentration (e.g., Bierman 1980). However, one recent modification to the structure of algal growth dynamics is the inclusion of variable nutrient stoichiometry for phytoplankton biomass. In their model application to Chesapeake Bay,

Cerco and Cole (1993) included a function which permitted variation in the carbon to phosphorus (C:P) ratio for phytoplankton biomass. This empirical function is developed from observed C:P ratios and soluble reactive phosphorus data in the upper bay. HydroQual (1995) recently implemented a more process-based function for variable carbon to nutrient stoichiometry, which also included variable carbon to chlorophyll stoichiometry, in a modeling study of the Massachusetts Bays system. This model formulation is based on a model developed by Laws and Chalup (1990). A unique feature of this model is that it accounts for variable carbon to chlorophyll ratios in phytoplankton due to both light and nutrient status.

Perhaps the two most significant changes that have been incorporated into modern eutrophication models are: (1) the addition of a coupled water column/sediment nutrient flux submodel and (2) the direct coupling of time-variable three-dimensional hydrodynamic and water quality models. The sediment nutrient flux model framework accounts for the deposition of organic matter from the water column to the sediment bed of the water body, its subsequent diagenesis or decomposition, and the flux of resulting end-products back to the overlying water column.

The model includes sediment processes for temperature and oxygen-dependent nitrification-denitrification, sorption of dissolved inorganic phosphorus and dissolved inorganic silica to sediment solids and the sorption of dissolved inorganic phosphorus to iron hydroxides in the aerobic layer of the sediment. The model also considers the generation of sediment oxygen demand, hydrogen sulfide and methane from the reduction of organic matter.

The model is developed and calibrated in Chesapeake Bay to an extensive multi-year data set (Di Toro and Fitzpatrick 1993). The model has been further verified against a wide-range of nutrient conditions using an extensive nutrient flux data set obtained from the University of Rhode Island MERL mesocosms. The model successfully reproduced the observed sediment nutrient composition and nutrient flux data using essentially the same parameter set as was used for Chesapeake Bay.

The only parameters to be changed between the two calibrations are the temperature-correction coefficients associated with diagenesis and the aerobic/anaerobic partition coefficients for phosphorus sorption. The MERL data were collected more frequently than the Chesapeake Bay data and, therefore, included more "cold weather" measurements than did the Chesapeake Bay data set. These additional data required minor adjustments to these temperature coefficients in order to reproduce the "cold weather" data. It also appeared reasonable that the phosphorus partition coefficients might be different between the two systems given differences in the iron content of the sediments in Chesapeake Bay and the MERL mesocosms (Narragansett Bay).

Examples of the coupling of water quality models to high resolution time-variable three-dimensional hydrodynamic models include the Chesapeake Bay system (Cerco and Cole 1993), the Massachusetts Bays system (HydroQual 1995, Blumberg and Fitzpatrick 1999). The use of well calibrated hydrodynamic models to drive water quality model computations is that it removes a degree of freedom in model calibration, i.e., the external specification of advective and dispersive transport. In addition, the use of high resolution hydrodynamic/water quality models removes, to a large degree, the "numerical errors" or numerical dispersion associated with coarse grid or box water quality models.

We believe that coupling eutrophication models to fisheries models becomes more speculative, although some efforts have been initiated in Chesapeake Bay and the Great Lakes community. While coupled hydrodynamic/water quality models of eutrophication can provide information to fisheries models, the projections of walleye production and fish community structure and diversity are made more complicated by the interactions between various predator/prey fish species, the impacts of overfishing and perhaps the relatively long life-spans of the fish themselves. Perhaps the area requiring the greatest research and monitoring effort is the area of the fishery.

The purpose of this chapter is to provide a history of the development and application of mathematical models of the eutrophication processes in Lake ecosystems and to comment on the potential use of eutrophication models in an ecosystem approach. However, it would be remiss if we did not include in this chapter, the research and development efforts conducted in the lakes of the Great Lakes ecosystem.

References

Anderson D.M. 1994. Red tides. *Scientific American* 271:62-68.

Angold P. G. 1997. The Impact of a Road Upon Adjacent Heathland Vegetation: Effects on Plant Species Composition. *The Journal of Applied Ecology* 34:409-417.

APIS. 2005. Website: Air Pollution Information System Eutrophication

Bartram, J., Wayne W. Carmichael, Ingrid Chorus, Gary Jones,is not hat easy and Olav M. Skulberg. 1999. Chapter 1. Introduction, In: *Toxic Cyanobacteria in Water: A guide to their public health consequences, monitoring and management.* World Health Organization.

Bertness M.D., P.J. Ewanchuk, and B.R. Silliman. 2002. Anthropogenic modification of New England salt marsh landscapes. *Ecology* 99:1395-1398.

Bertness M.D., P.J. Ewanchuk, and B.R. Silliman. 2001. Anthropogenic modification of Bierman, V.J., Jr., 1980. A comparison of models developed for phosphorus management in the Great Lakes. *Phosphorus Management Strategies for Lakes.* Ed. R.C. Loehr, C.S. Martin, W. Rast. Ann Arbor Science Publishers Inc., Ann Arbor, MI

Blumberg, A.F. and J.J. Fitzpatrick, 1999. Invited lecture: Strategy for coupling hydrodynamic and water quality models for addressing ling time scale environmental impacts. In *Environmental Hydraulics.* Ed. J.H.W. Lee, A.W. Jayawardena, and Z.Y. Wang. Proceedings of the second international symposium on environmental hydraulics. Hong Kong, China. A.A. Balkema, Rotterdam, Netherlands.

Buol S. W. 1995. Sustainability of Soil Use. *Annual Review of Ecology and Systematics* 26:25-44.

Carpenter, S.R., N.F. Caraco, and V.H. Smith. 1998. Non-point pollution of surface waters with phosphorus and nitrogen. *Ecological Applications* 8:559-568.

CERCO, C.F., and T. Cole, 1993. Application of the three-dimensional eutrophication model CE-QUAL-ICM to Chesapeake Bay. U.S. Army Engineer Waterways Experiment Station, Vicksburg, MS.

Cole J.J., B.L. Peierls, N.F. Caraco, and M.L. Pace. (1993). Nitrogen loading of rivers as a human-driven process. Pages 141-157 In M.J. McDonnell and S.T.A. Pickett, editors. *Humans as components of ecosystems.* Springer-Verlag, New York, USA.

Di Toro, D.M., D.J. O'Connor, J.L. Mancini and R.V.Thomann, 1973. A preliminary phytoplankton-zooplankton-nutrient model of Western Lake Erie. In *Systems Analysis & Simulation in Ecology.* 13. Academic Press.

Di Toro, D.M. and J.J. Fitzpatrick, 1993. Chesapeake Bay sediment nutrient flux model. Contract report EL-93-2. Prepared for U.S. Army Engineer Waterways Experiment Station, Vicksburg, MS

Gedney, R.T., 1971. Numerical calculation of the wind-driven currents in Lake Erie. Ph.D. Thesis, Department of Fluid, Thermal and Aerospace Sciences, Case Western Reserve University.

Herdendorf, C.E., 1967. Lake Erie physical limnology cruise, midsummer 1967. State of Ohio, Department of Natural Resources, Division of Geological Survey, Columbus, Ohio.

Hydroscience, Inc., 1973. Limnological systems analysis of the Great Lakes: Part I. Prepared for the Great Lakes Basin Commission. Hydroscience, Inc., Westwood, NJ.

HydroQual, Inc., 1995. A water quality model for Massachusetts and Cape Cod Bays: Calibration of the Bays Eutrophication Model (BEM). Prepared for the Massachusetts Water Resources Authority by HydroQual, Inc., Mahwah, NJ and Normandeau Associates, Bedford, NH.

Horrigan, L.; R. S. Lawrence, and P. Walker 2002. "How sustainable agriculture can address the environmental and human health harms of industrial agriculture". *Environmental Health Perspectives* 110: 445-456.

Hornung M., Sutton M.A. and Wilson R.B. [Eds.] 1995: Mapping and modelling of critical loads for nitrogen - a workshop report. Grange-over-Sands, Cumbria, UK. UN-ECE Convention on Long Range Transboundary Air Pollution, Working Group for Effects, 24-26 October 1994. Published by: Institute of Terrestrial Ecology, Edinburgh, UK.

Howarth R.W., G. Billen, D. Swaney, A. Townsend, N. Jaworski, K. Lajtha, J.A. Downing, R. Elmgren, N. Caraco, T. Jordan, F. Berendse, J. Freney, V. Kudeyarov, P. Murdoch, and Zhu Zhao-liang. 1996. Regional nitrogen budgets and riverine inputs of N and P for the drainages to the North Atlantic Ocean: natural and human influences. *Biogeochemistry* 35:75-139.

Huang W. Y., Y. C. Lu, and N. D. Uri. 2001. An assessment of soil nitrogen testing considering the carry-over effect. Applied Mathematical Modelling 25:843-860.

ILEC/Lake Biwa Research Institute [Eds]. 1988-1993 Survey of the State of the World's Lakes. Volumes I-IV. International Lake Environment Committee, Otsu and United Nations Environment Programme, Nairobi.

Kumazawa K. 2002. Nitrogen fertilization and nitrate pollution in groundwater in Japan: Present status and measures for sustainable agriculture. *Nutrient Cycling in Agroecosystems* 63:129-137.

Laws, E.A. and M.S. Chalup, 1990. A microalgal growth model. Limnol. *Oceanogr.* 35:597-608.

Lawton, L.A.; G.A. Codd 1991. "Cyanobacterial (blue-green algae) toxins and their significance in UK and European waters". *Journal of Soil and Water Conservation* 40: 87-97.

Martin, A.; G.D. Cooke (1994). "Health risks in eutrophic water supplies". *Lake Line* 14:

Mungall C. and D.J. McLaren. 1991. Planet under stress: the challenge of global change. Oxford University Press, New York, USA. 24-26.

New England salt marsh landscapes. *Ecology* 95:1395-1398.

O'Brien, J.W. 1974. The dynamics of nutrient limitation of phytoplankton algae: A model reconsidered. *Ecology* 55, 135-141.

Paerl H. W. 1997. Coastal Eutrophication and Harmful Algal Blooms: Importance of Atmospheric Deposition and Groundwater as "New" Nitrogen and Other Nutrient Sources. *Limnology and Oceanography* 42:1154-1165.

Raike A., O.P. Pietilainen, S. Rekolainen, P. Kauppila, H. Pitkanen, J. Niemi, A. Raateland, J. Vuorenmaa. 2003. Trends of phosphorus, nitrogen, and chlorophyll *a* concentrations in Finnish rivers and lakes in 1975-2000. *The Science of the Total Environment* 310:47-59.

Rodhe, W. 1969 Crystallization of eutrophication concepts in North Europe. In: *Eutrophication, Causes, Consequences, Correctives.* National Academy of Sciences, Washington D.C.

Smith, V.H.; G.D. Tilman, and J.C. Nekola 1999. "Eutrophication: impacts of excess nutrient inputs on freshwater, marine, and terrestrial ecosystems". *Environmental Pollution* 100: 179-196.

Shumway, S.E. 1990. "A review of the effects of algal blooms on shellfish and aquaculture". *Journal of the World Aquaculture Society* 21: 65-104.

Sharpley A.N., T.C. Daniel, J.T. Sims, and D.H. Pote. 1996. Determining environmentally sound soil phosphorus levels. *Journal of Soil and Water Conservation* 51:160-166.

Wright, S., L.H. Tiffany, W.M. Tidd, 1955. Limnological survey of Western Lake Erie. U.S. Department of the Interior, Fish and Wildlife Service, Special Scientific Report: Fisheries No. 139, Washington D.C.

❑❑❑

Chapter 14

Oil Spill Modeling

14.1 Introduction

An oil spill (or slick) is the intentional or unintentional release of oil into the natural environment as a result of human activity. The term often refers to marine oil spills where oil is released into the ocean or coastal waters (Fig. 14.1).

Fig. 14.1. Oil Pollution in Coastal Area

Oil can refer to many different materials, including crude oil, refined petroleum products or by-products, oily refuse, oil mixed in waste, or oily ballast. Oil is also released into the environment from natural geologic seeps on the seafloor, as along the coastline. The fate, behavior, and environmental effects of spilled oil can vary depending upon the type and amount of material spilled. In general, lighter refined petroleum products such as diesel and gasoline are more likely to mix in the water column and are more toxic to marine life, but tend to evaporate relatively quickly and do not persist long in the environment. Heavier crude or fuel oil, while of less immediate toxicity, can remain on the water surface or stranded on the shoreline for much longer. Oil from the Exxon Valdez and Gulf War oil spills, while weathering over time, has persisted along the shoreline for years after the spill.

With vast numbers living and depending on coastal areas for fishing and tourism throughout the world the consequences are evermore frightening. Such possibilities have caused outcries for oil companies, ship owners, and ship builders to share the responsibility of preventing such a disaster.

While ships today are more equipped than ever before the risks are far greater due to the ever increasing number of ships. One step in preventing further oil disasters is through designing boats in a manner that will minimize the chances of a spill. This is being done as ships today are increasingly requiring double hulls due to environmental regulation. Another step in mitigating the possible effects of an oil spill is through being able to contain and effectively treat spilt oil. A new vital step being developed in preventing oil damage is through installations of systems made for Fast Oil Recovery (FOR) of oil from wrecked ships. New innovations for mitigating possible oil spill consequence have met, however, an uncompromising shipping industry which historically has shown only to comply environmentally when either required or economically beneficial.

14.2 Largest Oil Spills (Oil Spill Case Histories 1967-1991)

The greater oilspill disaster occur in the world ocean are given below:

1. Gulf War oil spill, \ Gulf, January 23 1991
2. Hawaiian Patriot, N Pacific, February 26, 1977
3. Othello, Tralhavet Bay, Sweden, March 20, 1970
4. Braer, Shetland Islands, UK, January 5, 1993

5. Argo Merchant, Nantucket Island, December 21, 1976
6. Prestige, Galicia, Spain, November 13, 2002
7. Aegean Sea, A Coruña, Spain, December 3, 1992
8. Sea Empress, Wales, UK, February 15, 1996
9. World Glory, South Africa, June 13, 1968
10. Corinthos, Delaware River, Marcus Hook, Pennsylvania, USA, January 31, 1975
11. Burmah Agate, Galveston Bay, Texas, USA, November 1, 1979
12. One oil well, Gulf of Jason Edumdes, , 1979
13. Exxon Valdez, Prince William Sound, Alaska, USA, March 24, 1989
14. Keo, Massachusetts, USA, November 5, 1969
15. Storage Tank, Sewaren, New Jersey, USA, November 4, 1969
16. Ekofisk oil field, North Sea, April 22, 1977
17. Erika, Bay of Biscay, December 12, 1999
18. Tasman Spirit, Karachi, Pakistan, July 28, 2003
19. Selendang Ayu, Unalaska Island, Alaska, USA, December 8, 2004
20. Kharg 5, In the Canery islands and the slick hit Western Sahara and Morocco, Africa, December 1989.
21. Jiyeh power station oil spill, Lebanon, July 14-15, 2006
22. Kirki oil spill, Australia, July 20, 1991
23. Oil field, Persian Gulf, February, 1983
24. Atlantic Empress and Aegean Captain collision, Trinidad and Tobago, July 19, 1979
25. Castillo de Bellver, off Cape Town, South Africa, August 6, 1983
26. Amoco Cadiz (BP/Amoco, USA) - France, March 16 1978
27. Torrey Canyon, South England, March 18, 1967
28. Sea Star, Gulf of Oman, December 19, 1972
29. Urquiola, A Corunna, Spain, May 12, 1976

14.3 Causes and Effects of Oil Spill in Seas and Oceans (Nelson-Smith, 1973)

It shown that the environmental damage caused by oil spills can be greater than is previously thought. It is now thought that the impacts to marine life can be evident at less than one part per billion petroleum hydrocarbons. The lighter fractions of petroleum, such as benzene and toluene, are more toxic, but are more volatile and evaporate quickly. Heavier components of crude oil, such as polynuclear aromatic hydrocarbons (PAHs) appear to cause the most damage; while they are less toxic, they persist in the environment much longer than volatile components.

The worst pollution disaster in Europe to date is the 1991 sinking of 'The Haven', an oil tanker loaded with 144,000 tonnes of Cock oil and owned by Troodos Shipping a company owned by Stelios Haji-Ioannou (of easy Jet fame). 5 people were killed and for the next 12 years the Mediteranean coast of Italy and France have been polluted especially around Genoa and St Tropez. Haji-Ioannou was prosecuted and charged with manslaughter and bribery.

14.4 Pollution Problems in Harbour Areas

Overall in the harbour area, poor water quality has impacted the harbour's resources. As a result a number of beneficial use impairments have been identified for the area. These include:

- Restrictions on fish consumption
- Degradation of fish and wildlife populations
- Restrictions on dredging activities
- Eutrophication or undesirable algae
- Loss of fish and wildlife habitat.

Historically, Harbour has had significant water pollution problems due to industrial discharges, primarily from fish and vegetable processing plants. Since the early 1970's nutrient and metal concentrations have increased, and are largely attributed to non-point sources. The major environmental concerns in the historical sources of PCBs, contaminated sediments, high phosphorus concentrations, poor water clarity, bacterial contamination, and habitat loss.

The following steps to be implemented by harbour authority:

- Defining the problem
- Planning implementation
- Implementing the actions, and
- Monitoring restoration of the environment

Recommended remedial actions to restore the above environmental conditions were selected through the process, which included consultation with the public, NGO, Consultant, harbour engineers, hydraulic engineers.

Major improvements in water and sediment quality have resulted from upgrades to treatment facilities at the vegetable and fish processing plant. Construction of a secondary sewage treatment plant servicing part of the inputs of nutrients and bacteria from leaky septic systems and smaller processing plants. Upstream non-point sources are now the major contributors of these substances. Regular maintenance dredging of the harbour for shipping removes sediments and sediment-bound contaminants from the harbour, thus reducing the potential for contaminant re-entry to the food web. However, water circulation is poor and natural recovery processes are limited within the harbour.

14.5 Oil Spill Modeling

This technique typically uses the output from hydrodynamic and/ or advection-diffusion models to predict particle movements in a Lagrangian manner. The flow regime is seeded with particles having defined properties (size, density, settling velocity, etc) and tracked as they move with the flow. This is a useful means of visualising flow patterns, particularly eddies and recirculation cells but can also be used to examine the movement of material away from particular activities such as dredging, dumping, outfalls, etc. By examining the statistics of how particles are re distributed in the model domain, it is also possible to extract some quantitative information.

The discharged or spilled material is considered as particles being advected with the surrounding water body and dispersed as a result of random processes including the dispersion caused by current, wind-induced turbulence and molecular diffusion. To each particle a corresponding mass is attached. This mass can change during the simulation as a result of decay and the particles can deposit with a constant settling velocity and re suspend again.

14.5.1 *Data Requirements*

In addition to the data requirements outlined for the hydrodynamic and advection/diffusion the following requirements are identified:

- Model Set-up
  - Characteristics of particle to be tracked
- Boundary conditions
  - Seaward and riverine suspended sediment concentrations
  - Location of sediment inputs
  - Oil characteristics
  - Oilspill timings
  - Volume of Spill
  - Area of the Spill
- Calibration and verification data
  - Suspended sediment concentrations
  - Historic bathymetric data

Particle Tracking methods are an important technique for simulating transport phenomena such as transport of pollutants in coastal waters. Particle tracking methods can quite accurately predict the pollutant transport in cases of steep concentration gradients after the pollutant has just entered into the water whereas conventional methods such as finite difference and finite volume methods may have difficulties. Since the computation time in a particle model increases linearly with the number of particles, this often forms a limiting factor.

Oil spill modeling is very useful to visualising flow patterns, particularly eddies and recirculation cells but can also be used to examine the movement of oil material away from particular activities such as dredging, dumping, outfalls, accidental oilspill, etc.

References

The World Almanac and Book of Facts, 2004

Oil Spill Case Histories 1967-1991, NOAA / Hazardous Materials and Response Division, Seattle WA, 1992

Nelson and Smith, 1973Oil Pollution and Marine Ecology, Elek Scientific, London, 1972; Plenum, New York.

❑❑❑

Chapter 15

Tsunami Modeling

15.1 Introduction

A tsunami is a series of waves created when a body of water, such as an ocean, is rapidly displaced. Earthquakes, mass movements above or below water, volcanic eruptions and other underwater explosions, landslides, large meteorite impacts and testing with nuclear weapons at sea all have the potential to generate a tsunami. The effects of a tsunami can range from unnoticeable to devastating. The term tsunami comes from the Japanese words meaning *harbor* ("tsu",) and *wave* ("nami",). The term is created by fishermen who returned to port to find the area surrounding their harbor devastated, although they had not been aware of any wave in the open water. Tsunamis are common throughout Japanese history; approximately 195 events in Japan have been recorded.

A tsunami has a much smaller amplitude offshore, and a very long wavelength (often hundreds of kilometers long), which is why they generally pass unnoticed at sea, forming only a passing "hump" in the ocean (Fig. 15.1). Tsunami have been historically referred to as *tidal waves* because as they approach land, they take on the characteristics of a violent onrushing tide rather than the sort of cresting waves that are formed by wind action upon the ocean (Iwan, W.D., 2006).

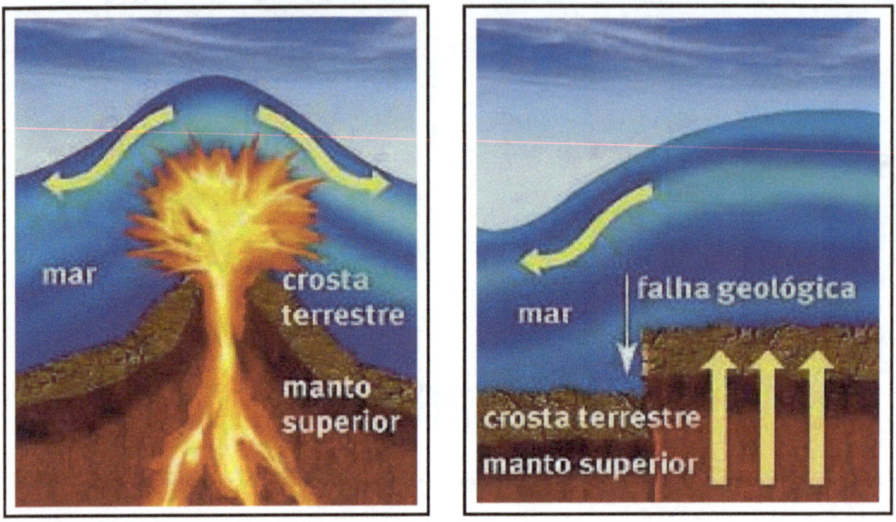

Fig. 15.1 Generation of Earthquake and Movement of Water

15.2 Causes and Impacts on Environment

A tsunami can be generated (Fig. 15.2) when the plate boundaries abruptly deform and vertically displace the overlying water. Such large vertical movements of the Earth's crust can occur at plate boundaries. Subduction earthquakes are particularly effective in generating tsunami. Also, one tsunami in the 1940's in Hilo, Hawaii, was actually caused by an earthquake on one of the Aleutian Islands in Alaska. That earthquake was measured with 7.8 on the Richter Scale.

Tsunami are formed as the displaced water mass moves under the influence of gravity and radiates across the ocean like ripples on a pond.

In the 1950s it was discovered that larger tsunami than previously believed possible could be caused by landslides, explosive volcanic action, and impact events when they contact water. These phenomena rapidly displace large volumes of water, as energy from falling debris or expansion is transferred to the water into which the debris falls. Tsunami caused by these mechanisms, unlike the ocean wide tsunami caused by some earthquakes, generally dissipate quickly and rarely affect coastlines distant from the source due to the small area of sea affected (Fig. 15.3).

An earth quake that measured 8.9 on the Richter Scale of the West Coast on Northern Sumatra set off a series of other earthquakes lasting 12 hours on the 26th of December (from 00:58 to 11:05 UTC), 2004 led to widespread catastrophe particularly in Sri Lanka, India, Maldives, Indonesia and Thailand with damage also in Malaysia, Bangladesh, Somalia and Seychelles (Fig. 15.5).

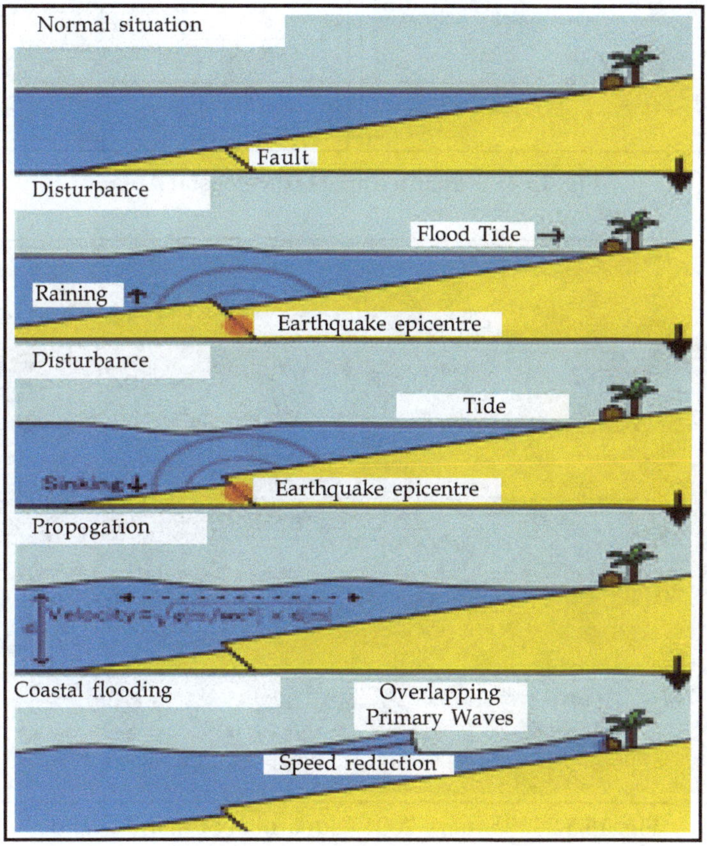

Fig. 15.2. Generation of Tsunami

Fig. 15.3. Tsunami Impacts on Coastal Area

Fig. 15.4. Tsunami Impacts on Coastal Area

Fig. 15.5 26th January 2004 Tsunami at Sumatra Island

These events can give rise to much larger local shock waves, such as the landslide at the head of Lituya Bay which produced a water wave estimated at 50 - 150 m and reached 524 m up local mountains. However, an extremely large landslide could generate a "megatsunami" that might have ocean-wide impacts. The geological record tells us that there have been massive tsunami in earth's past.

The effects of a tsunami (Fig. 15.4) can be mitigated by natural factors such as tree cover on the shoreline. Some locations in the path of the 2004 Indian Ocean tsunami escaped almost unscathed as a result of the tsunami's energy being sapped by a belt of trees such as coconut palms and mangroves. In one striking example, the village of

Naluvedapathy in India's Tamil Nadu region suffered minimal damage and few deaths as the wave broke up on a forest of 80,244 trees planted along the shoreline in 2002 in a bid to enter the Guinness Book of Records. Environmentalists have suggested tree planting along stretches of seacoast which are prone to tsunami risks. While it would take some years for the trees to grow to a useful size, such plantations could offer a much cheaper and long lasting means of tsunami mitigation than the costly and environmentally destructive method of erecting artificial barriers.

15.3 Modeling Tsunami

Information relating to the submarine earthquake in between Aceh, Indonesia and Sri Lanka of the 26th of December, 2004 has been discussed here. This compilation archives much of the readily available scientific information.

- The 9.0 Earthquake at 06.58 hours at the epicenter (and in Sri Lanka) led to a sequence of 15 other quakes across the Andaman region.
- While earthquakes could not be predicted in advance, once the earthquake was detected it would have been possible to give about 3 hours of notice of a potential Tsunami. Such a system of warnings is in place across the Pacific Ocean. However, there was no warning system in the Indian Ocean. In addition, coastal dwellers are educated in the Pacific littoral to get to high ground quickly following waves. However, those in the Indian Ocean were quite unaware.
- Tsunamis are rarer in the Indian Ocean as the seismic activity is much less than in the Pacific. However, there have been 7 records of tsunamis set off by earthquakes near Indonesia, Pakistan and one at Bay of Bengal.
- Earthquakes occur when any of the 12 or 13 plate collide at their boundaries. The present collision is due to compression between the Indian and Burmese plates. Scientists now believe that one plate that comprised the landmass from India to Australia has broken up into two. The initial 8.9 eruption happened near the location of the meeting point of the Australian, Indian and Burmese plates. Scientists have shown that this is a region of

compression as the Australian plate is rotating counterclockwise into the Indian plate. This also means that a region of seismic activity has become active in the South Eastern Indian Ocean.

- Tsunamis are not entirely unknown in Sri Lanka. For example, the tsunami in 1883 generated by the Volcanoes at Krakatoa led to a surge of at least 1 m in Sri Lanka. The damage was much less then. However, one difference was that this particular episode happened in the month of August. In the month of December, under the North-East monsoon, the Equatorial Indian Ocean jet propagates along the equator from Sumatra (near the epicenter of the quake) slightly to the South of Sri Lanka and to Somalia. This may be why the impact of the quake led to severe impacts in Sri Lanka.
- Once the large amount of pent up energy in the compression zones of the plate boundaries have been released, it takes another buildup of energy for another event of similar magnitude. This is unlikely in the short-term. However, in the future, Indian Ocean littoral regions should generate and pay attention to earthquake and tsunami warnings and be aware of the interplay of the seasonal oceanographic currents.

15.4 Observations and Model

After the 21 May 2003 Algiers earthquake, sea disturbances have been observed on the southeastern coast of the Majorca and Menorca Islands, as well as in Ibiza (Baleares). Witnesses have reported waves up to 2 m high, and a mean period of 10-12 min in Majorca. Tide gage records are available in Palma showing a maximum water height of about 60 cm.

These sea level variations can be related to a tsunami that could have been triggered either by the coseismic seafloor deformation induced by the submarine fault involved in the earthquake, or by a submarine landslide. In a first attempt we test here a source due to a coseismic seafloor deformation, that can be estimated through a model of elastic dislocation (Okada, 1985). This initial seafloor deformation is computed with the following parameters taken from Harward CMT solution. The mean displacement on the fault plane is about 85 cm. The first slip distributions obtained from seismological inversions show that the depth was probably shallower than the one we used here, therefore the initial seafloor displacement is probably larger.

Fault dimensions (km x km)	40 x 20
Strike, dip, rake	56°, 46°, 71°
Seismic moment (Nm)	$0.201\ 10^{20}$
Rigidity (Nm^2)	$30\ 10^{9}$
Epicenter	3.52°E - 36.98°N
Depth (km)	17

The initial deformation is assumed to be fully and instantaneously transmitted to the sea surface, where, through restoring gravity forces, tsunami waves begin to propagate across the sea. Since tsunami wavelengths are much larger than the mean water depth, we assume that the long wave theory is valid. The simulation is performed using a finite difference method that solves the hydrodynamic equations of continuity and of motion, including non linear terms. Nested bathymetric grids are used to account for the wave amplification near the coasts. Here we used data taken from the global bathymetry data deduced from altimetry (Smith and Wessel, 1997), and we build a 2' grid describing the western Mediterranean Sea, a 30″ grid for the Baleares Islands, and 10″ grids for the Majorca and Menorca Islands.

The modeling carried out here are very preliminary, and use rough bathymetric data only, especially for the fine grids. To account for the wave amplification reported along the SE coasts of Majorca or Menorca, fine data should be added in the bathymetric grids used for the simulation. Anyway these modelings show that a coseismic source can explain tsunami observations in the Baleares Islands, especially along their SE coasts. It is to be noted that tsunami observations along the Algerian coast have not be reported so far, and attention must be paid to future witnesses accounts. Finally, it is also worth mentioning that a dense and good quality tide gage network on the Mediterranean coasts could allow the implementation of inversion techniques to recover the submarine coseismic deformation, as it is performed in Japan for earthquakes located in the Nankai Trough (Tanioka and Satake, 2001 ; Baba *et al.*, 2002).

15.5 A Case Study – Tsunami of Chennai Coastal area

Vertical scales are not in true proportions in the figure, the total horizontal extent of the figure is 120 km, while the maximum elevation of the sea bottom and surface equal 2.4 m and 1.8 m respectively. The sketch is based on Okada's formula applied to the parameters of the

1969 earthquake, magnitude 7.9, southeast of Algarve, Portugal, near the Azores-Gibraltar fracture zone.

In May 2005, an international, interdisciplinary team of scientists conducted a detailed survey of the seafloor in the vicinity of the epicenter of the Great Sumatra earthquake of December 26, 2004-dubbed Sumatra Earthquake and Tsunami Offshore Survey (SEATOS 2005). The overall goal of the survey is to gather data to improve models of seafloor displacement and the resulting tsunami waves. To accomplish this goal, SEATOS built upon the results from the February 2005 HMS Scott multibeam data. Seven sites are selected from the HMS Scott data for SEATOS detailed investigations: a large underwater landslide, initially though to be the result of the December 26th earthquake: a ~20km long trench feature at the deformation front that was interpreted to represent recent seafloor displacement; seafloor locations on the three sites on the forearc high, and a steep mound north of Sumatra, interpreted as volcanic.

Operations begin after a 30-hour steam to the northeastern edge of the margin offshore Sumatra where high resolution seismic gear was deployed to collect the cross margin seismc data. Following this transect, detailed surveys including seismic, and Remotely Operated Vehicles (ROV) were conducted at each site. ROV surveys included high definition television quality video imaging of the seafloor, as well as core sampling for biological and geotechnical analyses.

SEATOS 2005 collected seismic data, ROV imagery, biological core data and sedimentological core data at sites along the accretionary complex deformation front, at one site in the forearc fold regions, and at sites along the forearc high. Early results suggest seafloor disturbances (from the Dec. 26th 2004 earthquake) and the formation of new seafloor features occurred at the deformation front and stable seafloor conditions prevailed at the forearc high. In addition, on the return to port, SEATOS investigated an unusual seafloor feature 200 m high, north of Sumatra that is a volcano, that is most likely part of the island arc. These data are currently being analyzed, along with recently published seismic inversion analyses, to develop a seafloor displacement model that best represents the Dec. 26th event.

During SEATOS, tsunami simulations are conducted for the Bay of Bengal using a wave model based on fully nonlinear Boussinesq equations. Such a model better represents the oscillatory wave trains that are typically observed during the event. Initially the model used the bathymetry and tsunami sources are defined using information from seismic inversion and hydroacoustic models. The comparison of model results with observations at tide gages, one satellite track, and runup measured in recent field surveys is quite good. A more detailed bathymetry is used to conduct case studies near the Northern tip of Sumatra and the coast of Thailand. These simulations are still in progress.

Tsunamis caused by earthquakes have long been viewed as one of the leading coastal hazards for seismic areas, such as the Pacific Ocean Basin. Many studies of the generation, propagation, and run up of such tsunamis have been conducted and extensively reported on in the literature. (Liu *et al.*,1991;Yeh *et al.*,1996). In view of recent research, tsunamis caused by Submarine Mass Failure (SMF),usually triggered, but not always, by seismic activity, also appear to represent another very significant local hazard for coastal areas.

For such SMF tsunamis, coastal run ups may be extremely large over small areas, although the energy associated both with the triggering earthquake and the associated SMF tsunami may be smaller than for tsunamis directly generated by large earthquakes.This results from various factors, such as the proximity to shore and the intense and more directional wave generation that may occur at the source, just above the SMF. Thus, some recent studies show that SMF tsunamis may be responsible for peak run up n about 30% of all measurable tsunamis (Watts and Borrero,2001).Clear indications of the repeated occurrence of SMF tsunamis over geologic times have been found all over the world. Many past events have been studied (Murty,1979) and modeled, both in the Laboratory (Wiegel,1955; Iwasaki, 1982; Heinrich,1992) and, more recently, using a variety of numerical simulation methods e.g.,Iwasaki, 1987; Harbitz,1992;Jiang and LeBlond,1992,1993,1994;Verriere and Lenoir, 1992). However, more than any past event, the 1998 Papua New Guinea (PNG) tsunami, and the high loss of life caused, spurred a renewed interest n SMF tsunamis n general, and the development of new SMF/landslide tsunami models n particular (e.g., Grilli and Watts,1999;Grilli *et al.*, 2002;Ward,2001).Some of these numerical studies, together with careful field surveys and geological studies of the PNG site, have led to very comprehensive analyses of what may have caused and happened during the PNG event (Kawata *et al.*,1999;Tappin *et al.*,1999;Tappin *et al.*,2001).In response to these and other studies, recent work by marine geologists now considers the tsunami potential of SMF scars (Goldfinger *et al.*,2000;Driscoll *et al.*,2000).

Earlier studies indicate that, after initial wave generation near the SMF, wave propagation rapidly becomes de coupled from the tsunami source (Jiang and LeBlond,1992). Hence, tsunami propagation and run up can be studied and modeled with the well-established methods/models used for other types of tsunamis. Note, however, that our recent experience indicates that the shorter SMF tsunami waves may require dispersive long wave equations, such as Boussinesqs, for a more accurate modeling of their propagation (Watts *et al.*, 2002b). Hence, the study of SMF tsunamis should specifically focus on: (i) the triggering mechanism, i.e., under which conditions and where will a SMF likely cause a tsunami to occur; (ii) the SMF mechanism & motion; and (iii) the generation of water waves by the SMF near the source.

Part (i) although key to hazard assessment in specific cases, relates to purely geological/geotechnical issues and will not be discussed here. Parts (ii) and (iii) can be viewed as two sides of a coupled problem, i.e. the SMF generates surface waves, while moving, as a result of its motion. A wide variety of numerical methods incorporating many different assumptions have been proposed to model SMF tsunami generation and early propagation (parts (ii) and (iii) above): (i) linear potential theory, with wave generation simulated by moving the domain boundary (Iwasaki,1987,1997;Verriere and Lenoir, 1992) ; (ii) linear superposition of constant velocity and constant depth Green's functions (Ward, 2001); (iii) depth-averaged Nonlinear Shallow Water (NSW)wave equations in combination with various landslide models (Fine *et al.*, 1998;Harbitz,1992;Imamura and Gica,1996;J ang and LeBlond,1992,1993,1994);(iv)fully nonlinear fluid dynamic field equations, in concert with assorted landslide models (Assier Rzadkiewicz *et al.*,1997; Grilli andWatts,1999; Grilli *et al.*,2002;Heinrich 1992).

To simplify the modeling of part (ii) above, SMFs have often been idealized and represented by solid blocks moving over plane slopes, in both laboratory e.g., Wiegel, 1955; Watts,1997 ; Watts, 2000 ; Watts *et al.*,2000,2002a;Raichlen and Synolakis,2002) and numerical studies (e.g., Grilli *et al.*, 1999, 2002). Others have used heavier fluids (i.e., mud flows)to simulate SMFs n their models (e.g.,Immamura and Gica,1996).Watts *et al.*(2000)compared these two approaches for three benchmark cases using models (for one case they also performed laboratory experiments)and showed, provided center of mass motions are similar, maximum wave amplitudes at the source were similar, although details of wave shape differed.

Watts (1997)and Watts and Grilli (2002)compared tsunami generation for different center of mass motions and different rates of underwater landslide and slump deformation and analyzed the sensitivity of maximum initial tsunami amplitude to geometric and landslide parameters, for idealized situations.

In light of earlier work, it appears that modelers have studied vastly different SMF geometries, motions, and constitutive behaviors. This makes a comparison of model results difficult. There is currently no real consensus in the community on the ability of these different models to reproduce SMF tsunami generation by underwater landslides. Moreover, many of these models have been developed in isolation from marine geology information, and therefore may not be particularly or easily applicable to real events. Even before considering their applicability to field problems, most of these techniques and ideas remain to be validated experimentally. Even the best field case studies, such as for the 1998 PNG event, have been plagued by difficulties and controversies despite the large amount of data available (Tappin *et al.*,2001).

The comparison of SMF tsunami generation models with experimental data for idealized benchmark problems appears to be a good and well proven way to evaluate and criticize the state of the art. Benchmark problems would permit to validate SMF tsunami numerical models and to help explain the origins of any discrepancies that may exist, both between numerical models and with comparisons to experimental results. Once they have been validated against a rigorous set of benchmark problems, numerical models presumably acquire some predictive capabilities, in the sense that they can be applied to similar problems with some degree of confidence. The tsunami community has faced similar challenges in the past, most recently with regard to tsunami propagation models. The NSF-funded 1995 Friday Harbor workshop provided experimental benchmark problems for tsunami propagation and runup models (Yeh *et al.*,1996). Tsunami modeling is to predict tsunami in the open coastal area to prevent damages in the human as well as materials.

References

Assier Rzadkiewicz,S.,Mariotti,C.,and Heinrich, P.(1997). "Numerical simulation of submarine landslides and their hydraulic effects. "*J. Wtrwy, Port, Coast, and Oc. Engrg.*, ASCE,123(4),149-157.

Driscoll, N.W., Weissel, J.K., Goff, J.A.(2000). "Potential for large-scale submarine slope failure and tsunami generation along the U.S.Mid-Atlantic *Coast.* "*Geology,* 28, 407-410.

Goldfinger, C., Kulm, L.D., McNeill, L.C., Watts, P.(2000)."Super-scale failure of the Southern Oregon Cascadia Margin."*Pure Appl.Geophys.,* 157, 1189-1226.

Fine, I.V., Rabinovich, A.B., Kulikov, E.A., Thomson, R.E., and Bornhold, B.D. (1998)."Numerical modelling of landslide-generated tsunamis with application to the Skagway Harbor tsunami of November 3,1994."*Proc., Tsunami Symp.,* Paris.

Grilli, S.T., and Watts, P.(1999)."Modeling of waves generated by a moving submerged body:Applications to underwater landslides."*Engrg.Analysis with Boundary Elements,* 23(8),645-656.

Grilli, S.T., Vogelmann, S., and Watts, P.(2002). "Development of a 3D numerical wave tank for modeling tsunami generation by underwater landslides."*Engrg.Analysis with Boundary Elements,* 23(8),645-656.

Harbitz, C.B. (1992). "Model simulations of tsunamis generated by the Storegga slides. "*Marine Geology,* 105, 1-21.

Heinrich, P.(1992). "Nonlinear water waves generated by submarine and aerial landslides. "J.Wtrwy, Port, *Coast, and Oc. Engrg.,* 118(3), 249-266.

Imamura, F., and Gica, E.C. (1996). "Numerical model for tsunami generation due to subaqueous landslide along a coast. "*Sci. Tsunami Hazards,* 14, 13-28.

Iwasaki, S. (1982). "Experimental study of a tsunami generated by a horizontal motion of a sloping bottom. "*Bull. Earth. Res. Inst.,* 57, 239-262.

Iwasaki, S.(1987). "On the estimation of a tsunami generated by a submarine landslide." *Proc., Int. Tsunami Symp.,* Vancouver, B.C., 134-138.

Iwasaki, S. (1997). "The wave forms and directivity of a tsunami generated by an earthquake and a landslide. "*Sci. Tsunami Hazards,* 15, 23-40.

Jiang, L., and LeBlond, P.H. (1992). "The coupling of a submarine slide and the surface waves which it generates. "*J. Geoph. Res.,* 97 (C8), 12731-12744.

Jiang, L., and LeBlond,P.H.(1993)."Numerical modeling of an underwater Bingham plastic mudslide and the waves which it generates. "*J. Geoph. Res.,* 98(C6), 10303-10317.

Jiang, L., and LeBlond, P.H. (1994). "Three-dimensional modeling of tsunami generation due to a submarine mudslide. "*J. Phys. Ocean.,* 24, 559-573.

Kawata, Y. and International Tsunami Survey Team members. (1999)."Tsunami in Papua New Guinea was intense as first thought. "*Eos, Trans. Am. Geophys. Union,* (abstract), 80(9), 101.

Liu, P.L.-F., Synolakis, C.E., and Yeh, H.(1991)."Report on the International Workshop on Long-Wave Run-up."*J.Fluid Mech.*, 229, 675-688.

Murty, T.S. (1979). Submarine slide-generated water waves n Kitimat Inlet, British Columbia. *J. Geoph. Res.*, 84 (C12), 7777-7779.

Raichlen, F., and Synolakis, C.E. (2002). "Waves and run-up generated by a three- dimensional sliding mass." Abstract, *European Geophysical Society Scientific Programme*, Nice, France .

Tappin, D.R., Matsumoto, T., and shipboard scientists.(1999). "Offshore surveys identify sediment slump as likely cause of devastating Papua New Guinea tsunami 1998." Eos, *Trans. Am. Geophys. Union*, 80(30), 329.

Tappin, D.R., Watts, P., McMurtry, G.M., Lafoy, Y., Matsumoto, T.(2001). "The Sissano, Papua New Guinea Tsunami of July 1998 —Offshore Evidence on the Source Mechanism."*Marine Geol.*, 175,1-23.

Verriere, M., and Lenoir, M.(1992)."Computation of waves generated by submarine landslides. "*Int.J.Num.Methods Fluids*, 14, 403-421.

Ward, S.N. (2001)."Landslide tsunami."*J. Geophys. Res.*, 106, 11201-11215.

Watts, P.(1997). "Water waves generated by underwater landslides,"PhD Thesis, California Inst.of Technol., Pasadena, CA.

Watts, P.(1998). "Wavemaker curves for tsunamis generated by underwater landslides. " J.Wtrwy, Port, *Coast and Oc. Engrg.*, ASCE, 124(3), 127-137.

Watts, P.(2000). "Tsunami features of solid block underwater landslides. "J. Wtrwy, Port, *Coast and Oc. Engrg.*, ASCE, 126(3), 144-152.

Watts, P., Imamura, F., and Grilli, S.T.(2000). "Comparing model s mulations of three benchmark tsunami generation cases. "*Sci.Tsunami Hazards*, 18(2),107-124.

Watts, P., and Borrero, J. C. (2001)."Probabilistic predictions of underwater landslides and tsunami amplitudes. "*Proc. of the Int. Tsunami Symp., Seattle,* WA, 697-710.

Watts, P.(2002). "The need for underwater landslide hazards prediction. " Sci. *Tsunami Hazards*, 20(2), 95-101.

Watts, P., and Grilli, S.T.(2002). "Tsunami generation by submarine mass failure I: Wavemaker models. "J. Waterways, Port, *Coast and Oc. Engng.*, submitted.

Watts, P., Imamura, F., Bengston, A., and Grilli, S.T. (2002a). "Benchmark Cases for Tsunamis Generated by Underwater Landslides. "*Proc. 4th Intl. Symp. on Ocean Wave Measurement and Analysis* (Waves 2001, San Francisco, CA)2, 1505-1514.

Watts, S.T. Grilli, J.T. Kirby, G.J. Fryer, D.R. Tappin (2002b) Simulation of landslide tsunami case studies using a Boussinesq model coupled to a fully non-linear tsunami source model. *Proc.EGS Conference,* Nice 4/02 (submitted).

Wiegel, R.L. (1955). "Laboratory studies of gravity waves generated by the movement of a submarine body. "*Trans. Am. Geophys. Union,* 36(5), 759-774.

Yeh, H., Liu, P.L.-F., and Synolakis, C.E.(1996). Long-wave run up models: Friday Harbor, USA, 12-17 Sept.1995. World Scientific, Singapore.

❑❑❑

Chapter 16

Littoral Drift Modeling

16.1 Introduction

Littoral drift is a geological process by which sediments such as sand or other materials, move along a beach shore.Long Shore Drift is the movement of eroded material, in a zig zag way, along the coast. The process arises when waves approach the shore obliquely. That is to say, at an angle other than 90°. The effect of this is determined by factors such as the direction and fetch of the present wind and, in the long term, of the prevailing wind.

Waves striking the shore at an angle as opposed to straight on will cause the wave *swash* to move up the beach at an angle. The swash moves the sediment particles up the beach at this angle, while the *backwash* brings them, solely under the influence of gravity, directly down the beach. Some geologists consider a beach to be merely a shoreline feature of deposited material, but William Bascom (1980) has argued that a beach is the entire system of sand set in motion by waves to a depth of ten meters or more off ocean coasts. Submerged, longshore bars are therefore also part of the beach, and thus beaches can be viewed as either:

- Small systems in which the rock material moves onshore, offshore, or alongshore by the forces of waves and currents; or
- Geological units of considerable size.

The former are described in detail below; the larger geological units are discussed elsewhere under bars. There are several conspicuous parts to a beach, all of which relate to the processes that form and shape it. The part mostly above water, and more or less actively influenced by the waves at some point in the tide, is termed the beach berm. The berm is the deposit of material comprising the active shoreline. The berm has a crest and a face the latter being the slope leading down towards the water from the crest. At the very bottom of the face, there may be a trough, and further seaward one or more longshore bars: slightly raised, underwater embankments formed where the waves first start to break.

The sand deposit may extend well inland from the *berm crest*, where there may be evidence of one or more older crests resulting from very large storm waves and beyond the influence of the normal waves. At some point the influence of the waves on the material comprising the beach stops, and if the particles are small enough, winds shape the feature. Where wind is the force distributing the grains inland, the deposit behind the beach becomes a dune.

The line between beach and dune is difficult to define in the field. Over any significant period of time, sand is always being exchanged between them. The drift line is one potential demarcation. This would be the point at which significant wind movement of sand could occur, since the normal waves do not wet the sand beyond this area. However, the drift line is likely to move inland under assault by storm waves.

16.2 Beach Formation

Beaches are deposition landforms, and are the result of wave action by which waves or currents move sand or other loose sediments of which the beach is made as these particles are held in suspension. Alternatively, sand may be moved by saltation. Beach materials come from erosion of rocks offshore, as well as from headland erosion and slumping producing deposits of scree. Some of the whitest sand in the world, along Florida's Emerald Coast, comes from the erosion of quartz in the Appalachian Mountains. A coral reef offshore is a significant source of sand particles.

The shape of a beach depends on whether or not the waves are constructive or destructive, and whether the material is sand or shingle. Constructive waves move material up the beach while destructive waves move the material down the beach. On sandy beaches, the backwash of the waves removes material forming a gently sloping beach. On shingle beaches the swash is dissipated because the large particle size allows percolation, so the backwash is not very powerful, and the beach remains steep. Cusps and horns form where incoming waves divide, depositing sand as horns and scouring out sand to form cusps. This forms the uneven face on some sand shorelines.

There are several beaches which are claimed to be the "World's longest", including Cox's Bazar, Bangladesh (120kms), Fraser Island beach, 90 Mile Beach in Australia and 90 Mile Beach in New Zealand and Long Beach, Washington. Wasaga Beach, Ontario on Georgian Bay claims to have the world's longest freshwater beach. But the longest beach in the world is in fact Praia do Cassino, a 240km long beach located in southern Brazil, near the border with Uruguay.

16.3 Beaches and Recreation

In the Victorian era, many popular beach resorts were equipped with bathing machines because even the all-covering beachwear of the period was considered immodest. This social standard still prevails in many Muslim countries. At the other end of the spectrum are clothing-optional beaches, where swimwear is optional.

A walk along the beach is also popular, including a long walk in the case of a long beach, for example from one seaside resort to the next. It is customary for people to walk barefoot to the beach, because of the pleasant feeling of sand on their soles and between their toes. The best beach walking areas typically are near the shoreline, where the sand is wet and more comfortable to walk in. A person will also enjoy walking with their bare feet in the water.

Due to intense use by the expanding human population, beaches are often dumping grounds for waste and litter, necessitating the use of beach cleaners and other cleanup projects. More significantly, many beaches are a discharge zone for untreated sewage in most underdeveloped countries; even in developed countries beach closure is an occasional circumstance due to sanitary sewer overflow. In these cases of marine discharge, waterborne disease from fecal pathogens and contamination of certain marine species is a frequent outcome.

This has the net effect of gradual movement of the particles *along* the shore by the use of swash and backwash. There is also erosion on the beach. The water molecules are among the particles moved in the process so there is a net movement of water as well. This establishes a longshore current.

Longshore drift (Fig. 16.1).is the name given to the process by which beach material is transported along the coast by the action of waves. Waves rarely hit the beach at exactly right angles to the coast, and are far more likely to hit the beach at an angle.

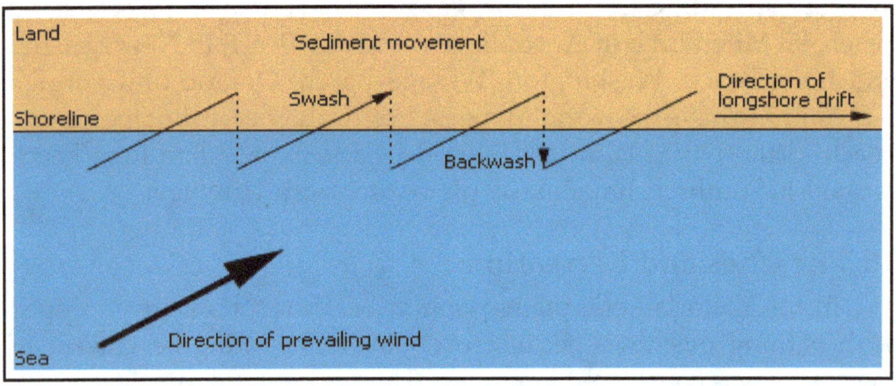

Fig. 16.1 Typical Longshore Drift in the Coastal Ecosystem

Where the wave front falls exactly squarely onto a major segment of coast, it will still meet parts of the shore in bays and estuaries obliquely, so that longshore drift will still occur on a more detailed scale.

The waves carry sediment from headlands into coves, as any seaside visitor will have noticed. Waves and currents can sort the pebbles, gravels, sands, and the muds to settle according to the energy remaining in the waves in the different locations.

16.4 Constructive Process

Littoral drift is one of the principal processes in the construction of spits, bars and tombolos (Fig. 16.2), and in the overall sustainability of beach deposits. Longshore currents and longshore drift are generally considered to be constructive processes. Unlike storm waves, they are not normally regarded as significant in coastal erosion, though the sedimentary particles have to come from somewhere. They are the continuing processes that nourish the beach and carry sand along the shore or a barrier spit to deposit it at the end of the spit so that the spit grows in length and size.

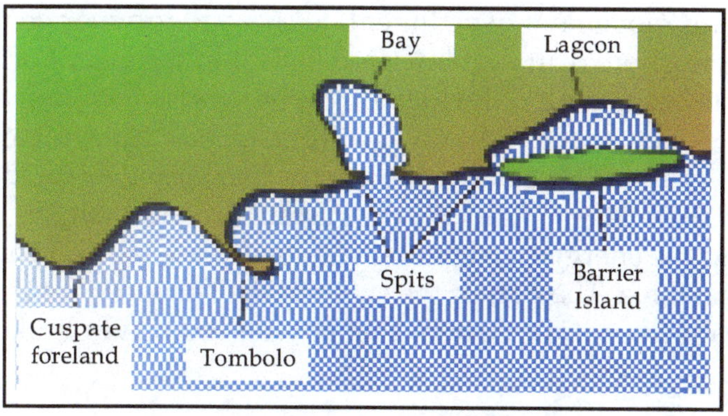

Fig. 16.2. Coastal Erosion and Accretion

The incident waves refracting into the entrance to the River and various beach forms produced in a variety of circumstances. The longshore drift from the east has brought material to the eastern side of the central groyne while the nearby reef has prevented deposition in quite the classical manner for such a position. On the leeward side of the groyne, inside the harbour entrance, small tombolos have formed. At whichever angle the wave front meets the general coast from time to time, it is diffracted into a constant direction as it enters the harbour so that on the wave-absorption beaches just inside, there are arcs of sediment ridges which the wave front always meets squarely. The sediment washes straight up the ridge and back.

The ebb tide is here, scouring the entrance and the silt carried by the harbour water is carried northwards. Close inshore there is an element of longshore current but most of this transport is being done by the tidal ebb stream which follows the coast northwards from three hours after high water to five hours before the next high water.

16.5 Destructive Process

In storm conditions particularly, the process can cause significant erosion to beaches and therefore is a threat to coastal communities that rely on their beaches for protection from more serious erosion. The need is most obvious where there are permanent structures on the shore. The loss of first the beach, then of old, glacially deposited, sediment to longshore drift resulted in the loss of the Tanushkodi town of Tamil Nadu in to the sea of Bay of Bangal.

The problem causes coastal towns in erosional longshore drift areas to employ coastal defences such as groynes, which interrupt the downstream movement of sediment. Jetties and groynes generally do not stop longshore drift, but interfere with it to the extent of retaining some of the beach material which might otherwise be removed. When a groyne is constructed, once equilibrium has been regained, the longshore drift resumes but, if the groyne has been well-sited, some of the beach material is retained on its upstream side. This is useful in dissipating the energy in incoming waves as the retained beach breaks some of the force. The part of the shore in the "shadow" of the groyne, downstream from it, is sheltered by having the waves broken by the structure.

A shoal forms and elongates at interruptions and terminations in littoral drift, such as at inlets and at islands in estuaries and bays. These features grow in the direction of the net littoral drift and are often manifested as subaerial spits. Waves approaching a shoreline at an angle transport sediment along the shore. Where the littoral drift reaches a break in the shoreline, sand is deposited, forming a submerged shoal or spit (Fig.16.3).

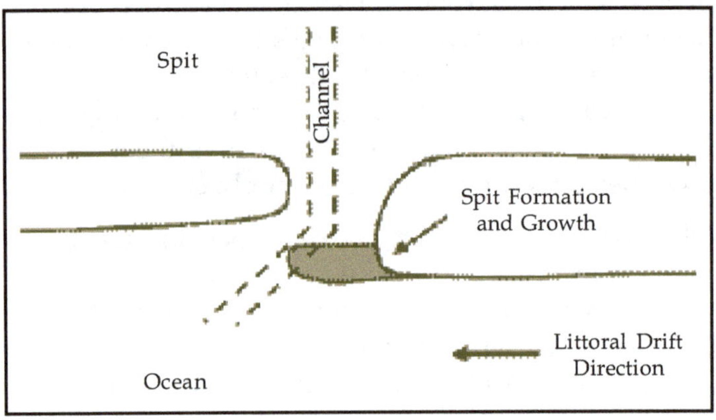

Fig. 16.3. Typical Littoral Drift in the Coastal Area

Littoral refers to the coast of an ocean or sea, or to the banks of a river, lake or estuary. It is usually used as an adjective, but may also be used as a noun. The littoral zone is defined as the area between the high water and low water marks. The word is derived from the Latin noun *litus, litoris,* meaning "shore. (Fig.16.4)

In lakes, where tides are usually negligibly small, other definitions of "littoral" must be used. The Minnesota Department of Natural Resources defines littoral as that portion of the lake that is less than 15 feet in depth. This zone is home to most of the aquatic plantlife (both rooted and floating) in a pond or lake because the high amount of sunlight reaching it allows for significant photosynthetic activity.

Fig. 16.4. Typical Erosion and Deposition in the Developed Jetties

The adjacency of water gives a number of distinctive characteristics to littoral regions. Water's erosive power results in particular types of landforms, such as sand dunes, and estuaries. The natural movement of the littoral along the coast is called the littoral drift. Biologically, the ready availability of water enables a greater variety of plant and animal life, and the additional local humidity due to evaporation usually creates a microclimate supporting unique types of organisms.

The littoral zone is bordered by the supra-littoral zone, also known as the "spray zone", and the sublittoral zone, which runs to the edge of the continental shelf.

The foreshore, also called the intertidal or littoral zone, is that part of a beach that is exposed by the low tides and submerged by high tides. This area can include many different types of habitats, including steep rocky cliffs, sandy beaches or vast mudflats. The area can be a narrow strip, as in Pacific islands that have only a narrow tidal range, or can include many meters of shoreline where shallow beach slope interacts with high tidal excursion.

16.6 Zonation

Marine biologists and others divide the intertidal region into three zones (low, middle, and high), based on the overall average exposure of the zone. The low intertidal zone, which borders on the shallow subtidal zone, is only exposed to air at the lowest of low tides and is primarily marine in character. The mid intertidal zone is regularily exposed and submerged by average tides.

The high intertidal zone is only covered by the highest of the high tides, and spends much of its time as terrestrial habitat. The high intertidal zone borders on the swash zone. On shores exposed to heavy wave action, the intertidal zone will be influenced by waves, as the spray from breaking waves will extend the intertidal region above the high tide line.

Depending on the substratum and topography of the shore, additional features may be noticed. On rocky shores, tide pools may be formed at low tide when water is trapped in hollows. Under certain conditions, such as those at Morecambe Bay, quicksand may be formed.

16.7 Ecology

The intertidal region is an important model systems for the study of ecology, especially on wave-swept rocky shores. The region contains a high diversity of species, and the different zones caused by the physics of the tides causes species ranges to be compressed into very narrow bands. This makes it relatively simple to study species across their entire cross shore range, something that can be extremely difficult in, for instance, terrestrial habitats that can stretch thousands of kilometers. Communities on wave swept shores also have high turnover due to disturbance, so it is possible to watch ecological succession over years rather than decades.

Since the foreshore is alternately covered by the sea and exposed to the air, organisms living in this environment must have adaptions for both wet and dry conditions. Hazards include being smashed or carried away by rough waves, exposure to dangerously high temperatures, and desiccation. Typical inhabitants include sea anemones, barnacles, chitons, crabs, isopods, limpets, mussels, starfish, snails and whales.

16.8 Definitions

A shoal forms and elongates at interruptions and terminations in littoral drift, such as at inlets and at islands in estuaries and bays. These features grow in the direction of the net littoral drift and are often manifested as subaerial spits.

16.9 Littoral Modeling

Predictions of the probable effects of proposed nearshore structures such as jetties, groins, intakes, and discharges on the longshore transport of sand require the establishment of the prevailing routes and rates of littoral drift. Littoral drift-the routes and rates of sand transport prior to the construction of the any structure in the coastal area. Nearshore structures operate as total or partial barriers to longshore drifted materials moving in that part of the littoral zone between the seaward end of the structure and the landward limit of wave uprush. Rarely are the nearshore circulation and sedimentation regimes of an area sufficiently well-known to allow realistic predictions of the changes a proposed nearshore structure will have on the prevailing sediment dispersal system. Commonly, intensive short term field studies of sediment transport are used to establish, as a basis for prediction, the existing littoral processes.

The predictions based on these data can be spurious if "typical" conditions are not encountered during the sampling period. A more reliable method of estimating the prevailing patterns of littoral drift in many areas is by inference from the long-term patterns of erosion and deposition. The results produced by the action of a set of littoral processes over a number of decades. This method utilizes "Nature's" role as a great averager, and eliminates some of the inconstancy which may characterize her over short periods of a few days or weeks. Fig.16.5 shows the erosion and deposition in port of Huron, Michigan.

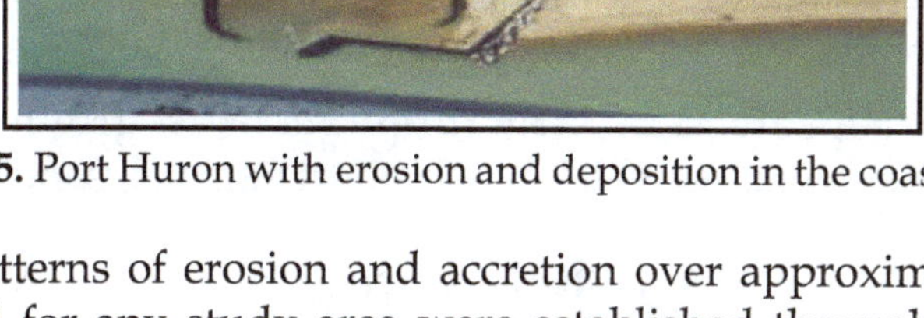

Fig. 16.5. Port Huron with erosion and deposition in the coastal area

The patterns of erosion and accretion over approximately a 100 year period for any study area were established through a detailed study of old and recent topographic survey sheets and original bathymetric sheets. These data were combined with a textural study of the cliff sediments and the sediments being deposited in the nearshore zone to produce a simple sand budget for this section of coast.

The nearshore circulation regime, consistent with the resulting patterns of erosion and accretion, is inferred from them. These data were then used to predict the probable effects proposed intake and discharge structures would have on the maintenance of this stretch of coast.

16.9.1 Steps involved in Littoral drift Modelling

- Calculate littoral drift at various sections along the coast.
  - This can be used to develop regional sediment budget for the coast.
- Determine dominant wave conditions contributing to the annual sediment transport at a given coast.
  - This can be done by calculating sediment transport for various events in the annual wave statistics.
- Determine "effective" depth – depth at which sediment transport is negligible at a given coast.
  - This can be obtained from the cross-shore distribution of sediment transport rates

16.10 Coastline Evolution Modeling

- Calculate long-term shoreline changes at a given coast using models such as LITPACK.
- The shoreline changes can be due to:
  - Natural effects due to alongshore changes in wave climate, shoreline orientation, beach profile *etc.*
  - New coastal developments such as harbours, ports, fishing horbours, dry and wet docs *etc.*
  - Shore protection structures such as groynes, jetties, breakwaters, seawalls, beach nourishments, sand mining *etc.*

Reference

Fisheries lake surveys. (2006)

❑❑❑

Chapter 17

An Ecohydrology Model

17.1. Introduction

Throughout human history, the coastal plains and Lowland River valleys have usually been the most populated areas over the world (Wolanski *et al.*, 2004). At present, about 60% of the world's population live along the estuaries and the coast (Lindeboom, 2002). This is degrading estuarine and coastal waters through pollution, eutrophication, increased turbidity, overfishing and habitat destruction. The pollutant supply does not just include nutrients; it also includes mud from eroded soil, heavy metals, radionuclides, hydrocarbons, and a number of chemicals including new synthetic products.

The impact on estuaries is commonly still ignored when dams and irrigation farming are proposed on rivers. In addition, estuaries are often regarded as sites for future development and expansion, and have been increasingly canalized and dyked for flood protection, and their wetlands infilled for residential areas.

All these factors impact on the biodiversity and productivity and hence, the overall health of estuaries and the ecosystem services they provide to humans. They increasingly lead humans away from the possibility of ecologically sustainable development of the coastal zone. Integrated coastal zone management plans are drawn up worldwide but, in the presence of significant river input, most are bound to fail because they commonly deal only with local, coastal issues, and do not consider the whole river catchment as the fundamental planning unit. It is as if the land, the river, the estuary, and the sea were not part of the same system. When dealing with estuaries and coastal waters, in most countries land-use managers, water-resources managers, and coastal and fisheries managers do not cooperate effectively due to administrative, economic and political constraints, and the absence of a forum where their ideas and approaches are shared and discussed (Wolanski *et al.*, 2004). To help alleviate this problem, UNESCO-IHP has launched the ecohydrology program. In this program, the concept of ecohydrology is introduced as a holistic approach to the management of rivers, estuaries and coastal zones within entire river catchments, by adopting science-based solutions to management issues that restore or enhance natural processes as well as the use of technological solutions (Zalewski, 2002).

This science based management requires the use of a model to quantify the human impact on the ecosystem health of estuaries and to enable the exchange of information between oceanographers, biologists, ecologists, engineers, sociologists, economists and water resources managers. This chapter describes such a model designed specifically for vertically well-mixed estuaries.

The model attempts to avoid being over-complex - the hydrodynamic sub-model is kept as simple as possible through they are still quite realistic. The ecological sub-model is also simple, though still realistic: it incorporates the six state variables. Thus the model is simpler than a number of other models (Findt and Kamp-Nielsen, 1997 - this comprises 12 state variables) that are often too complex and unwieldy for practical applications, especially when data are unavailable or insufficient.

17.2 The Science Behind the Model

The model is process based. The dominant physical, biological and human related processes in an estuary are assumed to follow those described by Wolanski *et al.* (2004). These processes are briefly summarised below.

The ecological health of estuaries is determined by the interaction between organisms and variations in salinity, currents, waves, suspended particulate matter (SPM), bed sediments, temperature, air exposure, hypoxia, wetland contaminants and biodiversity. Like the health of a living organism, the health of an estuary or a coastal water body, cannot be measured by one single variable, indeed a number of variables are important (Balls, 1994). Well-flushed estuaries are intrinsically more robust than poorly flushed systems. As a result, environmental degradation is most often apparent during periods of reduced freshwater inflows, *e.g.* during drought or when human activities reduce the freshwater flow.

Once riverine derived suspended particulate matter enters the estuary, it can be trapped within an Estuarine Turbidity Maximum (ETM) zone. The ETM is commonly located in the very low salinity reaches of an estuary. The maximum, depth-averaged, suspended solid concentration (SSC) at high water within an estuary can be predicted semi empirically as a function of the tidal length and the tidal range (Uncles *et al.*, 2002).

Sediment particles and aggregates within the ETM can give rise to marked changes in water quality. Fine particles can adsorb metal ions and organic macro molecules from solution to such an extent that some metals can be completely removed from solution within a strong ETM (Ackroyd *et al.*, 1986; Salomons and Forstner, 1984). Once nutrients enter an estuary, non-conservative behaviour can be pronounced. Key processes responsible for this non conservative behaviour include burial in sediment reservoirs and desorption processes particularly if the sediment is nutrient rich. Nutrients are generally mainly in particulate form in freshwater and can be released in solution in saline water.

The saltmarshes of Western Europe generally produce more than 1 kg m^{-2} yr^{-1} of above-ground dry matter (Boorman et al., 1994a, 1994b; Lefeuvre, 1996). Saltmarshes export some of this organic matter. Saltmarshes and their tidal creeks are also an important nursery ground and a refuge for larvae and postlarvae of bivalve, fish and zooplankton.

The estuary is modelled as a converter of living phytoplankton to detrital particles; it is also a conveyor of detrital matter to the sea. Fishes help transfer energy and matter from estuarine plants to upper trophic levels. The great bulk of the organic matter produced is processed through the detrital system. Zooplankton, planktivorous fish, interstitial micro and meiofauna, surface deposit-feeding molluscs, fishes and polychaeta, and filter-feeding invertebrates consume a much greater proportion of the primary production of the phytoplankton and benthic microalgae. Annual plant growth and decay provides continuing large quantities of organic detritus. In addition, there is often a considerable input of detritus from river inflow. Detrital particles and their associated microorganisms provide the basic food source for primary consumers such as zooplankton, most benthic invertebrates and some fishes. The first trophic level in the estuarine ecosystem is therefore best described as a mixed trophic level of detritus consumers, which in varying degrees are herbivores, omnivores or primary carnivores (Knox, 1986).

In the model, freshwater phytoplankton and bacterioplankton in the river are subject to salt stress when freshwater mixes with saltwater; and the freshwater microbial populations die in this zone (Flameling and Kromkamp, 1994; Goosen *et al.*, 1995). In the model, the salinity also limits the seaward distribution of saline water plankton, invertebrates and fishes.

17.3. The Estuarine Ecohydrology Model

The prototype is the Guadiana Estuary model at low flow conditions - because such low flow conditions now dominate since the Alqueva dam was constructed. For a freshwater flow Q_f less than 50 m^3s^{-1}, the Guadiana Estuary is vertically fairly well-mixed in salinity (Fortunato *et al.*, 2002). In a vertically well-mixed estuary, the distribution of a salinity S is determined from the solution of the 1-D advection-diffusion equation (Fischer *et al.*, 1979). The model enables one to readily calculate the flushing time of the estuary.

For a non-conservative constituent such as nutrients, plankton, detritus, fish, and bivalve, a sink-source term ΔC is included (Thomann, 1989), where C is the concentration, where ΔC is derived from the ecological sub-model described below.

The ecological sub-model is based on the non-linear Lotka-Volterra equation. The predator-prey relationship is calculated by non-linear equations (Kot, 2001; Brauer and Castillo-Chavez, 2001).

In an estuary changes in salinity constitute a major stress that can lead to death. There are other stressors, for instance small values of the dissolved oxygen concentration. A death rate must then be added to the equation for ΔC.

In an estuary, fringing wetlands can be an important source of detritus and nutrients, as well as a nursery for juveniles and sub-adults as well as a refuge. This is particularly the case for bivalves. Mathematically this is expressed by adding a source term to ΔC, based on the presence of intertial wetlands.

The ecosystem model represents mathematically the interactions between nutrients concentration (N), suspended sediment concentration (SSC), phytoplankton concentration (P), zooplankton concentration (Z), bivalve concentration (B), detritus concentration (D), zooplanktivorous fish concentration (ZF), and carnivorous/ omnivorous fish concentration (CF). All dying matter becomes detritus. Settling is not included in the model, because the animals (eg zooplankton) are mobile and can swim in the water. The model is equally complex at the lowest and highest trophic levels, which increases the model robustness (Jorgensen and Bendoricchio, 2001).

When using the zooplanktivorous fish equation, the discharge is modified to incorporate the horizontal swimming by the fish as fish swim, by kinesis or taxis following environmental clues (Wolanski *et al.*, 1997; Humston *et al.*, 2000).

17.4. Application to the Guadiana Estuary

The Guadiana River flows 810 km from its headwaters in Spain to its mouth in Portugal. The drainage basin area is 66,889 km^2, of which 17% is in Portugal and 83% in Spain. The mean yearly rainfall varies interannually between typically 400 to 600 mm. Summer months are usually dry and hot. The river flow varies seasonally, with freshet at 200–600 m^{3}s^{-1} and long periods of low flow usually in the range 10-20 m^{3}s^{-1} in summer.

The Guadiana Estuary is located in the southeast of Portugal, separating Portugal from Spain. The estuary is 60 km long, it has a maximum width of 550 m and the depth varies between 5 m and 17 m. The tidal regime of the estuary is meso-tidal, with average amplitude of 2 m. The Guadiana estuary is vertically well-mixed for river flows less than 20 m^{3}s^{-1} and is partially stratified up to a river discharge of 50 m^{3}s^{-1}. At higher river flows, a salt wedge forms. The estuary has large saltmarshes in its lower reaches. The parameter á varies along-channel to correspond to the location of the saltmarshes. The estuary supports the presence of species of economic interest, mainly fishes like the anchovy *Engraulis encrasicolus*. Moreover, the outwelling from the estuary to the coastal area promotes the development of the food web and influences the fisheries.

Several pollution sources exist in the Guadiana estuary area, mainly resulting from urbanization, agriculture (fertilizers, pesticides, herbicides), cattle breeding and olive oil production. The flow is at present regulated by more than 100 dams, including the Alqueve dam whose construction was completed in 2002 and that forms the largest man-made lake in Europe.

The model is successful of capturing the observations of along-river changes in salinity, nutrients, phytoplankton, zooplankton, bivalves, omnivorous fish and planktivorous fish in 2002 and 2003.

Further, a number of sensitivity tests were carried out. These model sensitivity tests are useful because they show that:

- The model appears robust because large but reasonable changes in the parameters do not lead to instabilities such as the destruction of trophic layers
- The biomass of organisms is directly affected by its consumption of prey or by its being predated on by predators higher in the food chain. Indirect effects across two trophic levels are generally small; for instance if we compare ZF from runs 1 and 5; ie there is no impact of the death rate of phytoplankton on carnivorous fish.
- Detritus is probably not important as a food source for zooplankton
- Salt marshes are most important for bivalves.
- Saltmarshes are important to prevent estuarine eutrophication.

17.5. Examples of Management Application of the Model

The model can readily be used to test scenarios that managers can ask, questions such as "what if?" when querying the impact of developments and disturbances to land use and water resources in the river catchment.

For instance, the model was used to answers the questions:

- What if the nutrient concentration in the Guadiana River is increased by a factor of two as a result of irrigation farming downstream of the Alqueve dam. Such farming is indeed planned.

- What if in addition the salt marshes were destroyed by developments?
- What is the influence of the Alqueve dam?
- What is the efficiency of various ecohydrology solutions to improve the ecological services provide by the estuary, such as creating wetlands creating freshets by releasing water from the Alqueve dam introducing new bivalve species in the freshwater part of the estuary removing nutrients from the river

Thus the estuarine ecohydrology model is able to provide answers to a number of practical questions. These answers must always be taken carefully because the model, like any ecosystem model, over-simplifies reality, and the data set is inadequate for a detailed calibration. In that sense the model predictions are somewhere between quantitative and qualitative. Detailed field studies are needed to better understand, and hence better parameterise in the model, the various processes driving the ecosystem. The model should then be seen as a living model - it has been written using subroutines that are readily edited, so that the new knowledge on individual processes can readily be incorporated in the model – the model is flexible can readily be modified.

For science, the model provides a tool to enable the exchange of information between oceanographers, biologists, ecologists, engineers, sociologists, economists and water resources managers.

It is hoped that the model can also be useful for management. The model shows that it is possible to predict within likely error bounds provided by the sensitivity tests the consequences on the estuary ecosystem health of human activities throughout the river catchment. The model does show that, to maintain the ecosystem services provided by the estuary, integrated coastal management needs to take the whole river catchment as the fundamental planning unit. It is necessary to bring together land-use managers, water-resources managers, and coastal and fisheries managers. The model offers thus a tool for use ecohydrology as a holistic approach to the management of rivers, estuaries and coastal zones within entire river catchments. The model can readily be used to assess future impact on the Guadiana Estuary ecosystem health caused by:

- Urbanisation or other factors that reduce the saltmarsh area
- Increase of nutrients load as result from changes in agriculture practices in the catchment area due to increase water availability by the Alqueva dam
- Extreme high freshwater discharges, ex due to release of high volume of water storage in the dam
- Introduction of exotic species.

The model can also be used to predict the efficiency of remedial measures, such as:

- Creating wetlands
- Creating freshets by releasing water from the Alqueva dam
- Managing introduced new bivalve species in the freshwater part of the estuary
- Removing nutrients from the river.

References

Ackroyd, D.R., Bale, A.J., Howland R.J.M, Knox, S., Millward, G.E. and Morris, A.W. 1986 Distributions and behaviour of dissolved Cu, Zn and Mn in the Tamar Estuary. *Estuarine, Coastal and Shelf Science* 23, 621-624.

Balls, P.W. 1994. Nutrient inputs to estuaries from nine Scottish East Coast rivers: influence of estuarine processes on inputs to the North Sea. *Estuarine, Coastal and Shelf Science* 39, 329-352.

Boorman, L.A., Hazelden, J.H., Loveland, P.J. and Wells, J.G. 1994a. Comparative relationships between primary productivity and organic and nutrient fluxes in four salt marshes. (In): Mitsch W.J. (ed.), *Global Wetlands. Old World and New*, pp. 181-189. Elsevier, Amsterdam.

Boorman, L.A., Hazelden, J., Andrews, R and Wells, J.G. 1994b. Organic and nutrient fluxes in four north-west European salt marshes. (In) Dyer, K.R. and Orth, R.J. (eds.). *Changes in Fluxes in Estuaries: Implications from Science to Management*, pp. 243-248. Olsen and Olsen, Fredensborg.

Brauer, F., and Castillo-Chavez, C. 2001. Mathematical models in population biology and epidemiology. Springer, Berlin, 416 pp.

Chícharo, M. A.; Chícharo, L..; Galvão, H.; Barbosa, A; Marques, M. H.; Andrade, J. P.; Esteves, E.; Miguel C. and Gouveia I. (2001). Status of the Guadiana estuary (South Portugal) during 1996-1998: *An ecohydrological approach. Aquatic Ecosystem Health and Management* 4, 73-90.

Chícharo L., Chícharo M.A. , Esteves E., Andrade P. and Morais P. 2002. Effects of alterations in fresh water supply on the abundance and distribution of Engraulis encrasicolus in the Guadiana Estuary and adjacent coastal areas of south Portugal. *Journal Ecohydrology and Hydrobiology* 1(3), 195-200

Doornbos, G. 1982 Changes in the fish fauna of the former Grevelingen estuary, before and after the closure in 1971. *Hydrobiological Bulletin* 16, 279-283.

Edwards, Andrew M. 2001 Adding Detritus to a Nutrient-Phytoplankton-Zooplankton Model: A Dynamical-Systems Approach. *Journal of Plankton Research* 23 (4), 389-413.

Flameling, I.A. and Kromkamp, J. 1994. Responses of respiration and photosynthesis of Scenedesmus protuberans (Fritsch) to gradual and steep salinity increases. *Journal of Plankton Research* 16, 1781-1791.

Flindt, M.R. and Kamp-Nielsen, L. 1997. Modelling of an estuarine eutrophication gradient. *Ecological modeling* 102, 143-153.

Fischer, H.B., List, E.Y., Koh, R.C.Y., Imberger, J. and Brooks, N.H. 1979. Mixing in inland and coastal waters. Academic Press, New York, 483 pp.

Fortunato, A.B., Oliviera, A. and Alves, E.T. 202. Circulation and salinity intrusion in the Guadiana Estuary. *Thalassas*, 18 (2), 43-65.

Goosen, N.K., van Rijswijk, P. and Brockmann, U. 1995. Comparison of hetereotrophic bacterial production rates in early spring in the turbid estuaries of the Scheldt and the Elbe. *Hydrobiologia* 311, 31-42

Hilborn, R., and Mangel, M. 1997. The ecological detective. Confronting models with data. Princeton University Press, Princeton,315 pp.

Humston, R., Ault, J.S., Lutcavage, M. and Olson, D.B. 2000. Schooling and migration of large pelagic fishes relative to environmental cues. *Fisheries Oceanography* 9 (2), 136-146.

Jorgensen, S.E. and Bendoricchio, G. 2001. Fundamentals of Ecological Modelling, Elsevier, 530 pp.

Kibirige, I. R Perissinotto, R., and Nozais, C. 2002. Alternative food sources of zooplankton in a temporarily-open estuary: evidence from ä^{13}C and ä^{15}N. *Journal of Plankton Research* 24 (10), 1089-1095.

Knox, G.A. 1986. Estuarine ecosystems: a systems approach (volume II). CRC Press, Florida. 230 pp.

Kot, M. 2001. Elements of mathematical ecology. Cambridge University Press, Cambridge, 464 pp.

Lefeuvre, J.C. 1996. Effects of Environmental Change on European Salt Marshes. Structure, functioning and exchange potentialities with marine coastal waters. Vols. 1-6. University of Rennes, France.

Lindeboom, H. 2002. The coastal zone: an ecosystem under pressure. In: Field, J.G., Hempel, G. and Summerhayes, C.P. (eds.), Oceans 2020: Science, trends, and the challenge of sustainability, pp. 49-84. Island Press, Washington, 365 pp.

Portela, L.I. 2004. Sediment delivery from the Guadiana estuary to the coastal ocean. *Journal of Coastal Research*, SI 39, Itajai, SC- Brasil.

Quinones, R.A., and Montes, R.M. 2001 Relationship between freshwater input to the coastal zone and historical landing of the benthic/demersal fish in Central-South Chile. *Fisheries Oceanography* 10, 311–328.

Roelke, D.L. 2000. Copepod food quality threshold as a mechanism influencing phytoplankton succession and accumulation of biomass, and secondary productivity: a modelling study with management implications. *Ecological modelling*, 134, 245-274

Salomons, W. and Forstner, U. 1984. Metals in the hydrocycle. Springer-Verlag, Berlin. 349 pp.

Simenstad, C.A., Morgan, C.A.; Cordell, J.R., Baross, J.A., 1994. Flux, passive retention, and active residence of zooplankton in Columbia River estuarine turbidity maxima. In: Dyer, J. and Orth, M. (Eds.). Changes in fluxes in estuaries: Implications from science to management. University of Plymouth. *ECSA22/ERF Symposium*. 473-482.

Simier, M., Blanc, L., Aliaume, C., Diouf, P.S. and Albaret, J.J. 2004 Spatial and temporal structure of fish assemblages in an "inverse estuary", the Sine Saloum system. *Estuarine, Coastal and Shelf Science* 59, 69–86.

Sklar, F. and Browder, J. 1998 Coastal environmental impacts brought about by alterations to freshwater flow in the Gulf of Mexico. *Environmental Management* 22, 547–562.

Thomann, R.V. 1980 Deterministic and statistical models of chemical fate in aquatic systems. pp. 322-351 in Levin, S.A., Hallam, T.G. and Gross, L.J. (editors) 'Applied Mathematical Ecology, Springer-Verlag, Berlin, 491 pp.

Uncles, R.J., Stephens, J.A. and Smith, R.E. 2002. The dependence of estuarine turbidity on tidal intrusion length, tidal range and residence time. *Continental Shelf Research* 22, 1835-1856.

Wolanski, E., Doherty,P. and Carleton, J. 1997. Directional swimming of fish larvae determines connectivity of fish populations on the Great Barrier Reef. *Naturwissenschaften*, 84, 6, 262-268

Wolanski, E., Boorman, L. A., Chícharo, L., Langlois-Saliou, E., Lara, R., Plater, A.J., Uncles, R.J., and Zalewski. M.. 2004. Ecohydrology as a new tool for sustainable management of estuaries and coastal waters. *Wetlands Ecology and Management*, 12, 235-276.

Zalewski, M. 2002. Ecohydrology—the use of ecological and hydrological processes for sustainable management of water resources. *Hydrological Sciences Bulletin* 47: 823-832.

□□□

Chapter 18

Commercially Available Models

18.1 Introduction

The commercially available water quality simulation models with various dimensions (1D, 2D and 3D) and database software designed to quantify movement and concentration of contaminants in lakes, streams, estuaries, and marine environments. The following water quality models were described below (Table 18.1).

18.2 Environmental Fluid Dynamics Code

The Environmental Fluid Dynamics Code (EFDC) is a multifunctional surface water modeling system, which includes hydrodynamic, sediment-contaminant, and eutrophication components. It has been applied to over 100 water bodies including rivers, lakes, reservoirs, wetlands, estuaries, and coastal ocean regions in support of environmental assessment and management and regulatory requirements. IT is a state-of-the-art hydrodynamic model that can be used to simulate aquatic systems in one, two, and three dimensions. It has evolved over the past two decades to become one of the most widely used and technically defensible hydrodynamic models in the world. It uses stretched or sigma vertical coordinates and Cartesian or curvilinear, orthogonal horizontal coordinates to represent the physical characteristics of a waterbody. It solves three-dimensional, vertically hydrostatic, free surface, turbulent averaged equations of motion for a variable-density fluid. Dynamically-coupled transport equations for turbulent kinetic energy, turbulent length scale, salinity and temperature are also solved. The model allows for drying and wetting in shallow areas by a mass conservation scheme.

Table 18.1. Commercially available water quality software (indicative)

Software Name	Operating System	Key Words
EFDC	Win 98, NT, 2000, XP	surface water, hydrodynamic, sediment-contaminant, eutrophication, rivers, lakes, reservoirs, wetlands, estuaries, coastal ocean regions, assessment, management, regulatory
EXAMS	Win 98, NT, 2000, XP	aquatic biology, assessment, biology, chemistry, compliance, environmental effects, metals, NPS related, permits, pesticides, point source(s), rivers, streams, surface water, test/analysis
EXPRESS	Win 98, NT, 2000, XP	przm, exams, exposure, shell
GCSOLAR	Win 9x, NT	solar, photolysis, half-life, pollutant, aquatic, ozone
HSCTM2D	Win 9x, NT	hydrology, sediment, contaminant, transport, finite element model, river, estuary
HSPF	Win 9x, NT	assessment, biology, compliance, deposition, discharge, environmental effects, estuaries, hydrology, lakes, metals, monitoring, NPS related, NPDES, nutrients, permits, pesticides, point source(s), rivers, sediment, streams, surface water, test/analysis, TMDL related, toxicity
HSPF Toolkit	Java enabled browser	HSPF, WinHSPF, FTABLE, RCHRES, BMP, assessment, compliance, discharge, environmental effects, hydrology, permits, rivers, sediment, streams, surface water, TMDL related, toxicity
PRZM3	Win 98, NT, 2000, XP	assessment, discharge, environmental effects, hydrology, land use management, metals, pesticides, surface water, test/analysis
QUAL2K	Win ME, 2000, XP	aquatic biology, assessment, compliance, discharge, environmental effects, hydrology, NPS related, NPDES, point source(s), surface water, test/analysis, TMDL related
SERAFM	MS-Excel 2003	serafm, exposure, assessment, mecury, hg, surface water, pond, stream, river
SWMM	Win 98, NT, ME, 2000, XP	aquatic biology, assessment, combined sewer, community, discharge, environmental effects, metals, NPS related, NPDES, point source(s), procedure, rivers, stormwater, streams, surface water, test/analysis, TMDL related

Contd...

Table 18.1. Contd...

Visual Plumes	Win 98, NT, 2000, XP	surface, water, jet, plume, model, quality, contaminant, TMDL
WASP	Win 95, 98, ME, 2000, XP	aquatic biology, assessment, compliance, discharge, environmental effects, hydrology, metals, NPS related, NPDES, point source(s), surface water, test/analysis, TMDL related
WHTIF	Win XP(SP2)	watershed, health, fish, habitat, macro-invertebrate, biodiversity, BASS, biomass, biota, aquatic ecosystem, MAHA, CVI, hydraulic, bankfull, flow, restoration, foodweb, mercury, PCB, dioxin
MIKE21	Win XP	Hydrodynamics, wave, water quality, temperature, BOD, DO, heavymetal, pollutants, ecology, sediment, flow, mud transport, littoral drift, coastline evolution etc.

EFDC can simulate water and water quality constituent transport in geometrically and dynamically complex water bodies, such as rivers, stratified estuaries, lakes, and coastal seas. The code solves the three-dimensional primitive variable vertically hydrostatic equations of motion for turbulent flow in a coordinate system which is curvilinear and orthogonal in the horizontal plane and stretched to follow bottom topography and free surface displacement in the vertical direction that is aligned with the gravitational vector. A second moment turbulence closure scheme relates turbulent viscosity and diffusivity to the turbulence intensity and a turbulence length scale. Transport equations for the turbulence intensity and length scale as well as transport equations for salinity, temperature, suspended cohesive and non-cohesive sediment, dissolved and adsorbed contaminants, and a dye tracer are also solved. An equation of state relates density to pressure, salinity, temperature and suspended sediment concentration.

The computational scheme utilizes an external-internal mode splitting to solve the horizontal momentum equations and the continuity equation on a staggered grid. The external mode, associated with barotropic long wave motion, is solved using a semi-implicit three time level scheme with a periodic two time level correction. A multi-color successive over relaxation scheme is used to solve the resulting system of equations for the free surface displacement. The internal mode, associated with vertical shear of the horizontal velocity components is solved using a fractional step scheme combining an implicit step for the vertical shear terms with an explicit step for all other terms. The transport equations for the turbulence intensity, turbulence length scale, salinity, temperature, suspended sediment, dissolved and adsorbed contaminants, and dye tracer are also solved using a fractional step scheme with implicit vertical diffusion and explicit advection and horizontal diffusion. A number of alternate advection schemes are implemented in the code.

Features of EFDC are its ability to simulate wetting and drying cycles, it includes a near field mixing zone model that is fully coupled with a far field transport of salinity, temperature, sediment, contaminant, and eutrophication variables. It also contains hydraulic structure representation, vegetative resistance, and Lagrangian particle tracking. EFDC accepts radiation stress fields from wave refraction-diffraction models, thus allowing the simulation of longshore currents and wave-induced sediment transport.

Applications and Possible Uses

- The EFDC model has been used for a study of high fresh water inflow events in the northern portion of the River and Lagoons.
- The model has been used for discharge dilution studies in the Rivers.
- Salinity intrusion studies include the River, Lagoon and Lake.
- Sediment transport studies include the River, Lake Bay,
- Power plant cooling studies include Reservoir, River and Bay.
- Contaminant transport and fate studies.
- Water quality eutrophication studies include Harbor, Bay, River Basin, Reservoir, The Bay water quality application is particularly noteworthy:
    - The model was calibrated using a one year data set and subsequently validated by simulation of an eight year historical period having extensive field data.
    - The model was then executed for 10 year management scenarios to develop a Comprehensive Conservation and Management Plan for the estuary system.

Model History

EFDC was originally developed at the Virginia Institute of Marine Science (VIMS) and School of Marine Science of The College of William and Mary, by Dr. John M. Hamrick. This activity was supported by the Commonwealth of Virginia through a special legislative research initiative. Subsequent support for EFDC development at VIMS was provided by the U.S. Environmental Protection Agency and the National Oceanic and Atmospheric Administration's Sea Grant Program. Tetra Tech, Inc. became the first commercial user of EFDC in the early 1990's and upon Dr. Hamrick's joining Tetra Tech in 1996, the primary location for the continued development of EFDC. Tetra Tech has provided considerable internal research and development support for EFDC over the past 10 years. Primary external support of both EFDC development and maintenance and applications at Tetra Tech has been provided by the U.S. Environmental Protection Agency including the Office of Science and Technology, the Office of Research and Development, and Regions 1 and 4. The ongoing evolution of the EFDC modeling system has to a great extent been application driven by a diverse group of EFDC users in the academic, governmental, and private sectors.

The EFDC model has been validated using analytical solutions, simulations of laboratory experiments and verified prototype applications. An extensive bibliography of referred journal and conference proceedings articles exist.

18.3 Exposure Analysis Modeling System (EXAMS)

The Exposure Analysis Modeling System (EXAMS) is an interactive software application for formulating aquatic ecosystem models and rapidly evaluating the fate, transport, and exposure concentrations of synthetic organic chemicals including pesticides, industrial materials, and leachates from disposal sites.

EXAMS contains an integrated Database Management System (DBMS) specifically designed for storage and management of project databases required by the software. User interaction is provided by a full-featured Command Line Interface (CLI), context-sensitive help menus, an on-line data dictionary and CLI users' guide, and plotting capabilities for review of output data. EXAMS provides 20 output tables that document the input data sets and provide integrated results summaries for aid in ecological risk assessments.

18.4 Express (Exams - PRZM Exposure Simulation Shell)

It is designed to facilitate a rapid and consistent assessment of aquatic pesticide exposure on a variety of crops. EXPRESS is a joint project of the USEPA Office of Pesticide Programs' Environmental Fate and Effects Division (OPP/EFED) and the USEPA Office of Research and Development (ORD).

18.5 GCSOLAR

GCSOLAR program is a set of routines that computes direct photolysis rates and half-lives of pollutants in the aquatic environment. The half-lives are calculated as a function of season, latitude, time-of-day, depth in water bodies, and ozone layer thickness.

This program operates in an interactive screen mode to facilitate data and program command entry by the user. Input values, with few restrictions, are format free. The user controls program flow by entering program execution commands. This release (1.10 February 1988 and 1.20 July 1999) of the GCSOLAR program differs from earlier versions of the program in that it permits the user to compute photolysis rate constants as a function of elevation above sea level. This is implemented with the GCSOLAR.

18.6 Hydrodynamic, Sediment, and Contaminant Transport Model (HSCTM-2D)

The Hydrodynamic, Sediment, and Contaminant Transport Model (HSCTM-2D) is a finite element modeling system for simulating two-dimensional, vertically-integrated, surface water flow (typically riverine or estuarine hydrodynamics), sediment transport, and contaminant transport.

The modeling system consists of two modules, one for hydrodynamic modeling (HYDRO2D) and the other for sediment and contaminant transport modeling (CS2D). One example problem is included. The HSCTM2D modeling system may be used to simulate both short term (less than 1 year) and long term scour and/or sedimentation rates and contaminant transport and fate in vertically well mixed bodies of water.

18.7 Hydrological Simulation Program - FORTRAN (HSPF)

Hydrological Simulation Program - Fortran (HSPF) is a comprehensive package for simulation of watershed hydrology and water quality for both conventional and toxic organic pollutants. HSPF incorporates watershed-scale ARM and NPS models into a basin-scale analysis framework that includes fate and transport in one dimensional stream channels. It is the only comprehensive model of watershed hydrology and water quality that allows the integrated simulation of land and soil contaminant runoff processes with In-stream hydraulic and sediment-chemical interactions. The result of this simulation is a time history of the runoff flow rate, sediment load, and nutrient and pesticide concentrations, along with a time history of water quantity and quality at any point in a watershed. HSPF simulates three sediment types (sand, silt, and clay) in addition to a single organic chemical and transformation products of that chemical.

18.8 PRZM Version Index

PRZM3 is the most recent version of a modeling system that links two subordinate models, PRZM and VADOFT, in order to predict pesticide transport and transformation down through the crop root and unsaturated zone.

PRZM is a one-dimensional, finite-difference model that accounts for pesticide and nitrogen fate in the crop root zone. PRZM3 includes modeling capabilities for such phenomena as soil temperature simulation, volatilization and vapor phase transport in soils, irrigation simulation, microbial transformation, and a method of characteristics (MOC) algorithm to eliminate numerical dispersion. PRZM is capable of simulating transport and transformation of the parent compound and as many as two daughter species. VADOFT is a one-dimensional, finite-element code that solves the Richard's equation for flow in the unsaturated zone. The user may make use of constitutive relationships between pressure, water content, and hydraulic conductivity to solve the flow equations. VADOFT may also simulate the fate of two parent and two daughter products. The PRZM and VADOFT codes are linked together with the aid of a flexible execution supervisor that allows the user to build loading models that are tailored to site specific situations. In order to perform probability-based exposure assessments, the code is also equipped with a Monte Carlo pre- and post processor.

The PRZM3 model system with documentation is available for microcomputer (DOS) systems. Enhancements to Release 3.0 include updates to physiochemical processes, increased flexibility in representing agronomic practices, and improved post-processing and data interpretation aids. A major modification includes algorithms that enable modeling of nitrogen cycle soil kinetic processes with the ability to track nitrogen discharges from a septic tank into the soil environment and movement to groundwater.

18.9 River and Stream Water Quality Model (QUAL2K)

QUAL2K (or Q2K) is a river and stream water quality model that is intended to represent a modernized version of the QUAL2E (or Q2E) model (Brown and Barnwell 1987). Q2K is similar to Q2E in the following respects:

One dimensional. The channel is well-mixed vertically and laterally.

- Steady state hydraulics. Non-uniform, steady flow is simulated.
- Diurnal heat budget. The heat budget and temperature are simulated as a function of meteorology on a diurnal time scale.
- Diurnal water-quality kinetics. All water quality variables are simulated on a diurnal time scale.
- Heat and mass inputs. Point and non-point loads and abstractions are simulated.

The QUAL2K framework includes the following new elements:

Software Environment and Interface. Q2K is implemented within the Microsoft Windows environment. It is programmed in the Windows macro language: Visual Basic for Applications (VBA). Excel is used as the graphical user interface.

- Model segmentation. Q2E segments the system into river reaches comprised of equally spaced elements. In contrast, Q2K uses unequally-spaced reaches. In addition, multiple loadings and abstractions can be input to any reach.
- Carbonaceous BOD speciation. Q2K uses two forms of carbonaceous BOD to represent organic carbon. These forms are a slowly oxidizing form (slow CBOD) and a rapidly oxidizing form (fast CBOD). In addition, non-living particulate organic matter (detritus) is simulated. This detrital material is composed of particulate carbon, nitrogen and phosphorus in a fixed stoichiometry.

- Anoxia. Q2K accommodates anoxia by reducing oxidation reactions to zero at low oxygen levels. In addition, denitrification is modeled as a first-order reaction that becomes pronounced at low oxygen concentrations.
- Sediment-water interactions. Sediment-water fluxes of dissolved oxygen and nutrients are simulated internally rather than being prescribed. That is, oxygen (SOD) and nutrient fluxes are simulated as a function of settling particulate organic matter, reactions within the sediments, and the concentrations of soluble forms in the overlying waters.
- Bottom algae. The model explicitly simulates attached bottom algae.
- Light extinction. Light extinction is calculated as a function of algae, detritus and inorganic solids.
- pH. Both alkalinity and total inorganic carbon are simulated. The river's pH is then simulated based on these two quantities.
- Pathogens. A generic pathogen is simulated. Pathogen removal is determined as a function of temperature, light, and settling.

18.10 Spreadsheet-Based Ecological Risk Assessment for the Fate of Mercury (SERAFM)

SERAFM Conceptual Model - SERAFM is a process-based, steady-state modeling system based in a spreadsheet framework incorporating a series of process modules. SERAFM predicts speciated mercury concentrations (Hg0, HgII, MeHg, HgT) in water (filtered and unfiltered) and sediments, and total mercury concentrations in fish tissue. Using this information exposure risk hazard indices are estimated for a series of exposed wildlife. Three scenarios of historical contamination from contaminated sediments, reference conditions, and feasible remediation concentrations are simultaneously and instantaneously predicted. SERAFM is not a priori predictive, but rather requires site-specific calibration. This is a challenge due to the limits of understanding in mercury process science, and is inherent in any process-based modeling structure. SERAFM was specifically developed to serve as a screening-level tool to investigate mercury cycling and exposure concentrations in water bodies and associated impacted wildlife exposure risk, as well as a research tool to investigate mercury processes within the watershed and water body.

SERAFM is a steady-state, process based mercury cycling model designed specifically to assist a risk assessor or researcher in estimating mercury concentrations in the water column, sediment, and fish tissue for a given water body for a specified watershed. SERAFM predicts mercury concentrations in these media for the species Hg0, HgII, and MeHg. The model runs three simultaneous scenarios. One scenario is for historically contaminated sediment, where the total mercury concentration in sediments is known. This scenario would be effective, for example, for modelling a Superfund site where the sediment is acting as a loading source to the system. In this first case, the total mercury concentration in the sediment is entered into the model as a known parameter. The second scenario is a hypothetical background or reference condition, which is defined as the condition as if no historical loading of mercury had occurred at this site. Therefore, the mercury concentrations in water and sediment are calculated with no known mercury sediment concentration, but rather the total mercury concentration sediment is directly calculated by the model. Mercury loadings to the water body come from atmospheric deposition and watershed erosion and runoff, and, in this scenario, the water body sediments act as a sink rather than a possible source to the system. Using the calculated results of these two scenarios, a third scenario is run as a proposed, possible sediment clean-up goal using a linear interpolation of the calculated sediment concentration and the hazard indices of the identified most sensitive species as a gauge for the proposed clean-up sediment concentration. Then, from this information, the concentrations of mercury in the water body and fish tissue are calculated, and wildlife and human hazard indices are calculated. The series of sub-modules within SERAFM include: mercury loading; abiotic and biotic solids balance, equilibrium partitioning; water body mercury processes; and wildlife risk calculations.

SERAFM is solved using a steady-state assumption, therefore all differential, governing equations are solved by setting the differential to zero. Being a CSTR model, there are no required boundary or initial conditions. To solve the system of linear equations a function was written using Visual Basic for Applications (VBA) in Microsoft Excel 2003©. This is required because Microsoft Excel cannot solve a system of simultaneous, coupled equations. The specific function, called LINEAR_SOLVE, uses L-U Decomposition to solve the given linear algebra equation: A*x=b, where A is an m x n matrix, x is an n x 1 matrix, and b is an m x 1 matrix. By using a VBA function, the model predictions are updated instantaneously whenever any parameter is changed. The function was compared to analytical solutions for several test cases to validate that the function was operating properly.

Applications and Possible Uses

- Prediction and evaluation of steady-state mercury cycling in an aquatic ecosystem consisting of a river reach or lake (stratified or well-mixed) and its associated watershed.
- Prediction of risk (hazard indices) for wildlife exposed to a water body with mercury contaminated sediments (such as historical industrial releases).
- Investigate impact of mercury deposition to a water body.
- Investigate sensitivity of processes and model parameters.

Model History

SERAFM was originally developed in response to ERASC Request #10 in 2003 to address developing a remediation goal for mercury in sediments for exposed wildlife, particularly at historically contaminated sites, such as at a Superfund Site. Additionally, the application of a proprietary process-model demonstrated the need for the US EPA's own mercury process-based water body model that allowed for transparency and ease-of-use. SERAFM is designed to be self-explanatory, flexible for use in different modeling scenarios, and self-contained with modules being housed in separate worksheets linked through an entire spreadsheet workbook.

SERAFM has been used in its steady state mode for modeling current scenario conditions of mercury dynamics in five ecosystems across the US, and adapted to evaluate the temporal response of changes in mercury concentrations in support of the Clean Air Mercury Rule.

18.11 Storm Water Management Model (SWMM)

The EPA Storm Water Management Model (SWMM) is a dynamic rainfall-runoff simulation model used for single event or long-term (continuous) simulation of runoff quantity and quality from primarily urban areas. The runoff component of SWMM operates on a collection of subcatchment areas that receive precipitation and generate runoff and pollutant loads. The routing portion of SWMM transports this runoff through a system of pipes, channels, storage/treatment devices, pumps, and regulators. SWMM tracks the quantity and quality of runoff generated within each subcatchment, and the flow rate, flow depth, and quality of water in each pipe and channel during a simulation period comprised of multiple time steps.

SWMM was first developed in 1971, and has since undergone several major upgrades since then. It continues to be widely used throughout the world for planning, analysis and design related to stormwater runoff, combined sewers, sanitary sewers, and other drainage systems in urban areas, with many applications in non-urban areas as well. The current edition, Version 5, is a complete re-write of the previous release. Running under Windows, SWMM 5 provides an integrated environment for editing study area input data, running hydrologic, hydraulic and water quality simulations, and viewing the results in a variety of formats. These include color-coded drainage area and conveyance system maps, time series graphs and tables, profile plots, and statistical frequency analyses.

Capabilities

SWMM accounts for various hydrologic processes that produce runoff from urban areas. These include:

- Time-varying rainfall
- Evaporation of standing surface water
- Snow accumulation and melting
- Rainfall interception from depression storage
- Infiltration of rainfall into unsaturated soil layers
- Percolation of infiltrated water into groundwater layers
- Interflow between groundwater and the drainage system
- Nonlinear reservoir routing of overland flow.

Spatial variability in all of these processes is achieved by dividing a study area into a collection of smaller, homogeneous subcatchment areas, each containing its own fraction of pervious and impervious sub-areas. Overland flow can be routed between sub-areas, between subcatchments, or between entry points of a drainage system.

SWMM also contains a flexible set of hydraulic modeling capabilities used to route runoff and external inflows through the drainage system network of pipes, channels, storage/treatment units and diversion structures. These include the ability to:

- Handle drainage networks of unlimited size
- Use a wide variety of standard closed and open conduit shapes as well as natural channels
- Model special elements such as storage/treatment units, flow dividers, pumps, weirs, and orifices
- Apply external flows and water quality inputs from surface runoff, groundwater interflow, rainfall-dependent infiltration/ inflow, dry weather sanitary flow, and user-defined inflows
- Utilize either kinematic wave or full dynamic wave flow routing methods
- Model various flow regimes, such as backwater, surcharging, reverse flow, and surface ponding
- Apply user-defined dynamic control rules to simulate the operation of pumps, orifice openings, and weir crest levels

In addition to modeling the generation and transport of runoff flows, SWMM can also estimate the production of pollutant loads associated with this runoff. The following processes can be modeled for any number of user-defined water quality constituents:

- Dry-weather pollutant buildup over different land uses
- Pollutant washoff from specific land uses during storm events
- Direct contribution of rainfall deposition
- Reduction in dry-weather buildup due to street cleaning
- Reduction in washoff load due to BMPs
- Entry of dry weather sanitary flows and user-specified external inflows at any point in the drainage system
- Routing of water quality constituents through the drainage system
- Reduction in constituent concentration through treatment in storage units or by natural processes in pipes and channels

Applications

Since its inception, SWMM has been used in thousands of sewer and stormwater studies throughout the world. Typical applications include:

- Design and sizing of drainage system components for flood control
- Sizing of detention facilities and their appurtenances for flood control and water quality protection
- Flood plain mapping of natural channel systems (SWMM 5 is a FEMA-approved model for NFPI studies)
- Designing control strategies for minimizing combined sewer overflows
- Evaluating the impact of inflow and infiltration on sanitary sewer overflows
- Generating non-point source pollutant loadings for waste load allocation studies
- Evaluating the effectiveness of BMPs for reducing wet weather pollutant loadings.

18.12 Visual Plumes

The Visual Plumes model system is a Windows-based software application for simulating surface water jets and plumes. It also assists in the preparation of mixing zone analyses, Total Maximum Daily Loads (TMDLs), and other water quality applications.

18.13 Water Quality Analysis Simulation Program (WASP)

The Water Quality Analysis Simulation Program. (WASP7), an enhancement of the original WASP (Di Toro *et al.*, 1983; Connolly and Winfield, 1984; Ambrose, R.B. *et al.*, 1988). This model helps users interpret and predict water quality responses to natural phenomena and manmade pollution for various pollution management decisions. WASP is a dynamic compartment-modeling program for aquatic systems, including both the water column and the underlying benthos. WASP allows the user to investigate 1, 2, and 3 dimensional systems, and a variety of pollutant types. The time varying processes of advection, dispersion, point and diffuse mass loading and boundary exchange are represented in the model. WASP also can be linked with hydrodynamic and sediment transport models that can provide flows, depths velocities, temperature, salinity and sediment fluxes.

WASP has been used to examine eutrophication of Tampa Bay, FL; phosphorus loading to Lake Okeechobee, FL; eutrophication of the Neuse River Estuary, NC; eutrophication Coosa River and Reservoirs, AL; PCB pollution of the Great Lakes, eutrophication of the Potomac Estuary, kepone pollution of the James River Estuary, volatile organic pollution of the Delaware Estuary, and heavy metal pollution of the Deep River, North Carolina, mercury in the Savannah River, GA.

WASP Preprocessor

The data preprocessor allows for the rapid development of input datasets. The ability to bring data into the model is as simple as cut and paste or queried from a database. The preprocessor provides detailed descriptions of all model parameters and kinetic constants. When linking WASP with hydrodynamic models it is as simple as pointing to the hydrodynamic linkage file.

- Import time series from WRDB, Spreadsheet, Text Files
- Automatically import hydrodynamic model interface information
- Multi-session capable
- Run time diagnosis

Post Processor

The Post-Processor (MOVEM) provides an efficient method for reviewing model predictions and comparing them with field data for calibration. MOVEM has the ability to display results from all of the WASP models as well as others. MOVEM allows the modeler to displays the results in two graphical formats:

1. *Spatial Grid :* a two dimensional rendition of the model network is displayed in a window where the model network is color shaded based upon the predicted concentration.
2. *x/y Plots :* generates an x/y line plot of predicted and/or observed model results in a window.

There is no limit on the number of x/y plots, spatial grids or even model result files the user can utilize in a session. Separate windows are created for each spatial grid or x/y plot created by the user.

18.14 WHATIF - Watershed Health Assessment Tools Investigating Fisheries

WHATIF is software that integrates a number of calculators, tools, and models for assessing the health of watersheds and streams with an emphasis on fish communities in the Mid-Atlantic Highland region. The tool set consists of hydrologic and stream geometry calculators, a fish assemblage predictor, a fish habitat suitability calculator, macro-invertebrate biodiversity calculators, and a process-based model to predict biomass dynamics of stream biota (the BASS model). The tools can be used to assess conditions and associated stressors in aquatic ecosystems, to examine causes of impairment, and to forecast ecological outcomes of habitat alterations and fisheries management actions. WHATIF also supports screening analysis, such as prioritization of areas for restoration and comparison of alternative watershed and habitat management scenarios. The toolkit was developed for the Mid-Atlantic Highlands region of the United States in order to support Canaan Valley Institute.

WHATIF is software that integrates a number of calculators, tools, and models for assessing the health of watersheds and streams with an emphasis on fish communities. The tool set consists of hydrologic and stream geometry calculators, a fish assemblage predictor, a fish habitat suitability calculator, macro invertebrate biodiversity calculators, and a process-based model to predict biomass dynamics of stream biota (the BASS model).

The software is built upon a database of stream measurements which serve as inputs to the calculators and tools. The calculators and tools vary in complexity, and can take the form of regression-based predictions up to a process based fish population model, the Bioaccumulation and Aquatic System Simulator (BASS). The toolkit allows the user to investigate a location on a stream or river by examining the data that have been collected at the location, and using the data with the calculators to compare actual conditions with predicted or possible conditions.

Or example, the Fish Assemblage Predictor might indicate that a stream in this location would likely support a trout population, while the data show that no trout have been observed. Then, the Habitat Suitability calculator could be used to examine the scores for trout to see if an explanation can be found in the current stream conditions. The user could explore what changes could be made to the stream, such as overhanging vegetation or erosion control, to allow trout to thrive. Finally, trout could be "stocked" in the simulation model to see if they persist and how other species are affected.

All of the data, both original observations and user-generated experimental data, are stored in the central database which allows access by any WHATIF tool or calculator. Thus, the output of one tool can become the input of another tool, and this feature is exploited by some calculators and tools with certain kinds of data.

Applications and Possible Uses

- Watershed health assessment.
- Prediction of stream hydraulic geometry based on mean annual flow conditions.
- Prediction of stream hydraulic geometry based on Bankfull flow conditions.
- Prediction of fish habitat suitability.
- Prediction of fish assemblage.
- Prediction of macro-invertebrate biodiversity (EPT and MBII).
- Simulation of bio-accumulation.
- Investigate sensitivity of processes and model parameters.
- Screening analysis, such as prioritization of areas for restoration and comparison of alternative watershed and habitat management scenarios.

WHATIF was developed in collaboration with the Canaan Valley Institute (CVI) to address aquatic ecosystem management and restoration needs of stakeholders in the Mid Atlantic Highlands region of the United States. WHATIF version 1 was tested by states and nonprofit groups at a CVI Fisheries Management Workshop (December 2005). Subsequent to this usability review, a number of new data, scientific and interface enhancements resulted in the current release, version 2. All WHATIF tools, including a new fisheries stocking, harvest and simplified chemical exposure tool from the Bioaccumulation and Aquatic System Simulator (BASS) model, were developed specifically for this effort. WHATIF version 2 supports both region-wide queries of aquatic ecosystem health and mapping of streams in the Mid Atlantic Highlands as well as stream reach-level investigations of habitat suitability, community dynamics and foodweb bioaccumulation for mercury, PCBs and dioxin-like compounds.

18.15 MIKE21

Water quality modelling is a numerical lab for Ecological Modelling. It is a generic and open tool for customizing aquatic ecosystem models to describe for instance water quality and eutrophication. It expertise and know how concerning ecological modelling has been collected in predefined ecosystem descriptions to be loaded and used in water quality software. So the water quality modelling templates describe physical, chemical and biological processes related to environmental problems and water pollution. The following is a common description of the WQ (water quality) templates:

1. WQSimpleTandSCOLI.ecolab
2. WQSimpleTandS.ecolab
3. WQSimpleCOLI.ecolab
4. WQSimple.ecolab
5. WQNutrients.ecolab

The templates describes the resulting concentrations a number of variables such as bacteria, which threatens bathing water quality, oxygen depletion due to the release of BOD, excess concentrations of nutrients, chlorophyll-nutrient interactions and degradation of chemical substances.

It is integrated with the advection-dispersion module, which describes the physical transport processes at each grid-point covering the area of interest. Other data required are concentrations at model boundaries, flow and concentrations from pollution sources, water temperature, etc.

The water quality modelling system solves the process equations using a rational extrapolation method in an integrated two-step procedure with the advection-dispersion module. Only the water quality processes are included in the description and equations explained below.

Applications

- The water quality module is used for a range of environmental investigations:
- Studies of hygienic problems related to bacteria from sewage outfalls and other sources.
- Survival of bacteria related to different environmental conditions.
- Oxygen conditions affected by BOD, ammonia and other oxygen consuming substances.
- Spreading, degradation and interaction between the inorganic nutrient: ammonia, nitrite, nitrate and phosphate.
- Evaluation of potential for eutrophication problems related to nutrient levels (nitrogen and phosphorus) and chlorophyll-a.
- Decay of chemical substances and effect evaluation based on resulting concentration levels.

Mathematical Formulations

The MIKE 21/3 ECOLab solves the system of differential equations describing the physical, chemical and biological interactions involved in the survival of bacteria, degradation of organic matter, resulting oxygen conditions and excess levels of nutrients in coastal areas.

The following variables can be modelled using DHI supported WQ templates:

Variable	Symbol
Dissolved BOD	BOD_d
Suspended BOD	BOD_s
Sedimentated BOD	BOD_b
Ammonia (NH_{4+}-N)	NH_3
Nitrite (NO_{2-}- N)	NO_2
Nitrate (NO_{3-}-N)	NO_3
Dissolved oxygen	DO
Phosphorus	PO_4
Faecal coliforms	CF
Total coliforms	CT
One or more user defined pollutants	(UDP_1...)

The processes and concentrations of the parameters are influences by external factors such as incident solar radiation (coliform bacteria decay) and discharges. Several combinations of the listed variables are implemented as "model levels" securing maximum flexibility. At the most simple level only BOD and DO are modelled and the most complex level includes all the variables.

Data Requirements

Basic Model Parameters

Model Grid size and extent

Time step and length of simulation

Type of output required and its frequency

Bathymetry and Hydrodynamic Input

Combined Advection-Dispersion Model

Dispersion coefficients

Initial Conditions

Concentration of parameters

Boundary Conditions

Concentration of parameters

Pollution Sources

Discharge magnitudes and concentration of parameters

Process rates

Size of coefficents governing the process rates

Only some of these coefficients can be determined by calibration. Others will be based on literature. The values or found from actual measurements and laboratory test.

The mathematical model runs were taken for large number of combinations of various input parameters in MIKE21 software. From the MIKE-21 model studies it was observed that the transport of pollutants are dispersed in x ,y and z directions with thick concentration to thin concentration based on the direction of currents, waves, tides.

The above list of Software, which are commercially available are mindicative.

❐❐❐

Terms and Definitions

Abatement : Reducing the degree or intensity of, or eliminating, pollution.

Abiotic : Not relating to living things, not alive.

Activated Carbon : A highly adsorbent form of carbon used to remove odors and toxic substances from liquid or gaseous emissions. In waste treatment it is used to remove dissolved organic matter from waste water. It is also used in motor vehicle evaporative control systems.

Acute toxicity : A chemical stimulus severe enough to rapidly induce an effect; in aquatic toxicity tests, an effect observed within 96 hours or less is considered acute. When referring to aquatic toxicology or human health, an acute effect is not always measured in terms of lethality.

Adsorption : The process by which a gas, vapor, dissolved material or very small particle adheres to the surface of a solid due to chemical or physical forces; the attraction and adhesion of ions from an aqueous solution to the solid soil or rock surfaces with which they are in contact.

Adsorption-desorption : Adsorption is the process by which nutrients such as inorganic phosphorous adhere to particles via a loose chemical bond with the surface of clay particles. Desorption is the process by which inorganic nutrients are released from the surface of particles back into solution.

Advection : The transport of dissolved contaminants by the bulk movement of water flow; the main process driving the movement of dissolved contaminants.

Aeration : A process of supplying or introducing air/oxygen into a medium which promotes biological degradation of organic matter in water. The process may be passive (as when waste is exposed to air), or active (as when a mixing or bubbling device introduces the air).

Aerobe : Bacteria that use oxygen as an electron acceptor. Life or processes that require, or are not destroyed by, the presence of oxygen.

Aerobic : Environmental conditions characterized by the presence of dissolved oxygen; used to describe biological or chemical processes that occur in the presence of oxygen.

Algae : Simple rootless plants that grow in sunlight, waters, on rocks and in soil, in proportion to the amount of available nutrients. They can affect water quality adversely by lowering the dissolved oxygen in the water. They are food for fish and small aquatic animals.

Alkaline, Alkalinity : The condition of water or soil, which contains a sufficient amount of alkali substances to raise the pH above 7.0. The capacity of water to neutralize acids.

Alluvial : Relating to mud and/or sand deposited by flowing water.

Ambient : Usual or natural surrounding conditions, e.g., ambient temperature-the natural, uninfluenced temperature of the surroundings.

Ambient water quality : Natural concentration of water quality constituents prior to mixing of either point or nonpoint source load of contaminants. Reference ambient concentration is used to indicate the concentration of a chemical that will not cause adverse impact to human health.

Ammonia : Inorganic form of nitrogen; product of hydrolysis of organic nitrogen and identification. Ammonia is preferentially used by phytoplankton over nitrate for uptake of inorganic nitrogen.

Ammonia toxicity : Under specific conditions of temperature and pH, the un ionized component of ammonia can be toxic to aquatic life. The un ionized component of ammonia increases with pH and temperature.

Anaerobic : A life or process that occurs in, or is not destroyed by, the absence of oxygen.

Assimilative capacity : The amount of contaminant load (expressed as mass per unit time) that can be discharged to a specific stream or river without exceeding water quality standards or criteria. Assimilative capacity is used to define the ability of a waterbody to naturally absorb and use waste matter and organic materials without impairing water quality or harming aquatic life.

Attenuation : The process by which a compound is reduced in concentration with distance and time through absorption, adsorption, degradation, dilution, diffusion, dispersion, and/or chemical or biological transformation.

Aquatic ecosystem : Complex of biotic and abiotic components of natural waters. The aquatic ecosystem is an ecological unit that includes the physical characteristics (such as flow or velocity and depth), the biological community of the water column and benthos, and the chemical characteristics such as dissolved solids, dissolved oxygen, and nutrients. Both living and nonliving components of the aquatic ecosystem interact and influence the properties and status of each component.

Bacteria : (Singular: bacterium) Microscopic living organisms ubiquitous in the environment, that can aid in pollution control by metabolizing organic matter in sewage, oil spills or other pollutants. However, bacteria in soil, water or air can also cause human, animal and plant health problems.

Background levels : Background levels represent the chemical, physical, and biological conditions that would result from natural geomorphological processes such as weathering or dissolution.

Bed Load : Sediment particles resting on or near the channel bottom that are pushed or rolled along by the flow of water.

Bedrock : Any solid rocks exposed at the surface or overlain by unconsolidated materials.

Benthic : Refers to material, especially sediment, at the bottom of an aquatic ecosystem.It can be used to describe the organisms that live on, or in, the bottom of a waterbody

Benthos : A form of aquatic plant or animal life that is found on or near the bottom of a stream, lake, ocean or other water body.

Benthic Region : The bottom layer of a body of water.

Best management practices (BMPs) : Methods, measures, or practices that are determined to be reasonable and cost effective means for a land owner to meet certain, generally nonpoint source, pollution control needs. BMPs include structural and nonstructural controls and operation and maintenance procedures.

Bias : Consistent deviation of measured values from the true value, caused by systematic errors in a procedure.

Biochemical oxygen demand (BOD) : The amount of oxygen per unit volume of water required to bacterially or chemically oxidize (stabilize) the oxidizable matter in water. Biochemical oxygen demand measurements are usually conducted over specific time intervals (5,10,20,30 days). The term BOD generally refers to standard 5 day BOD test.

Biomass : The amount, or weight, of a species, or group of biological organisms, within a specific volume or area of an ecosystem.

Boundary conditions : Values or functions representing the state of a system at its boundary limits.

Brackish : Mixed fresh and salt water.

Calibration : Testing and tuning of a model to a set of field data not used in the development of the model; also includes minimization of deviations between measured field conditions and output of a model by selecting appropriate model coefficients.

Channel : A natural stream that conveys water; a ditch or channel excavated for the flow of water

Chemical Oxygen Demand (COD) : A measure of the oxygen required to oxidize all compounds, both organic and inorganic, in water.

Chlorophyll : Green photosynthetic pigment present in many plant and some bacterial cells. There are seven known types of chlorophyll; their presence and abundance vary from one group of photosynthetic organisms to another.

Chloride : An atom of chlorine in solution, bearing a single negative charge

Coastal Zone : As defined by the NCP, all US waters subject to the tide, US waters of the Great Lakes, specified ports and harbors on inland rivers, waters of the contiguous zone, other waters of the high seas subject to the NCP, and the land surface or land substrata, groundwaters, and ambient air proximal to those waters. The term coastal zone delineates an area of federal responsibility for response action. Precise boundaries are determined by EPA/USCG agreements and identified in federal regional contingency plans.

Coastal Waters : Those waters surrounding the continent which exert a measurable influence on uses of the land and on its ecology. The Great Lakes and the waters to the edge of the continental shelf.

Complete mixing : No significant difference in concentration of a pollutant exists across the transect of the waterbody.

Concentration : Amount of a substance or material in a given unit volume of solution. Usually measured in milligrams per liter (mg/l) or parts per million (ppm).

Conservation : Preserving and renewing, when possible, human and natural resources. The use, protection, and improvement of natural resources according to principles that will assure their highest economic or social benefits.

Conservative substance : Substance that does not undergo any chemical or biological transformation or degradation in a given ecosystem.

Contaminant - 1) Any physical, chemical, biological, or radiological substance or matter that has an adverse affect on air, water, or soil.

Contamination : Act of polluting or making impure; any indication of chemical, sedimentt, or biological impurities.

Conventional pollutants : As specified under the Clean Water Act, conventional contaminants include suspended solids, coliform bacteria, biochemical oxygen demand, pH, and oil and grease.

Decay : Gradual decrease in the amount of a given substance in a given system due to various sink processes including chemical and biological transformation, dissipation to other environmental media, or deposition into storage areas.

Decay Constant : A constant which expresses the probability that an atom or molecule of a chemical will decay in a given time interval.

Decomposition : Metabolic breakdown of organic materials; the by products formation releases energy and simple organics and inorganic compounds. (see also respiration)

Design Capacity : The average daily flow that a treatment plant or other facility is designed to accommodate.

Designated use : Uses specified in water quality standards for each waterbody or segment regardless of actual attainment.

Diagenesis : Production of sediment fluxes as a result of the flux of particulate organic carbon in the sediment and its decomposition. The diagenesis reaction can be thought of as producing oxygen equivalents released by various reduced species.

Dilution : Addition of less concentrated liquid (water) that results in a decrease in the original concentration.

Diurnal : (1) Occurring during a 24-hr period; diurnal variation. (2) Occurring during the day time (as opposed to night time). (3) In tidal hydraulics, having a period or cycle of approximately one tidal day.

Discharge : 1) Flow of surface water in a stream or canal or the outflow of groundwater from a flowing artesian well, ditch, or spring. 2) Discharge of liquid effluent from a facility or of chemical emissions into the air through designated venting mechanisms.

Detritus : Any loose material produced directly from disintegration processes. Organic detritus consists of material resulting from the decomposition of dead organic remains.

Dissolved Oxygen (DO) : The oxygen freely available in water, vital to fish and other aquatic life and for the prevention of odors. DO levels are considered a very important indicator of a water body's ability to support desirable aquatic life. Secondary and advanced waste treatment are generally designed to ensure adequate DO in waste-receiving waters.

Dissolved Solids : Disintegrated organic and inorganic material in water. Excessive amounts make water unfit to drink or use in industrial processes. Generally noticeable in concentrations greater than 500 mg/L.

Drought : In general, an extended period of dry weather, or a period of deficient rainfall that may extend over an indefinite number of days, without any quantitative standard by which to determine the degree of deficiency needed to constitute a drought. Qualitatively, it may be defined by its effects as a dry period sufficient in length and severity to cause at least partial crop failure or impair the ability to meet a normal water demand.

Dynamic model : A mathematical formulation describing the physical behavior of a system or a process and its temporal variability.

Dye study : Use of conservative substances to assess the physical behavior of a natural system to given stimulus

Estuary : That portion of a coastal stream influenced by the tide of the body of water into which it flows; a bay, at the mouth of a river, where the tide meets the river current; an area where fresh and marine water mix

Ecosystem : An interactive system that includes the organisms of a natural community association together with their abiotic physical, chemical, and geochemical environment.

Effective Porosity for Flow (n_e or n_{ef}) : Represents the interconnected porosity of a material. In a porous material, some void spaces may be saturated but are not able to transmit water, and so are not available for flow. These spaces represent dead zones of immobile water. n_{ef} corrects the porosity (n) of a material to account for these dead spaces.

Effluent : Municipal sewage or industrial liquid waste (untreated, partially treated, or completely treated) that flows out of a treatment plant, septic system, pipe, etc.

Estuary : 1) A semi-enclosed coastal body of water that has a free connection with the open sea and within which seawater is measurably diluted with fresh water from land drainage. 2 See Wetlands. ater or sewage.

Eutrophication : Enrichment of an aquatic ecosystem with nutrients (nitrates, phosphates) that accelerate biological productivity (growth of algae and weeds) and an undesirable accumulation of algal biomass.

Eutrophication model : Mathematical formulation that describes the advection, dispersion, and biological, chemical, and geochemical reactions that influence the growth and accumulation of algae in aquatic ecosystems. Models of eutrophication typically include one or more species groups of algae, inorganic and organic nutrients (N,P), organic carbon, and dissolved oxygen.

Extinction coefficient : Measure for the reduction (absorption) of light intensity within a water column.

Flocculation : The process by which suspended colloidal or very fine particles are assembled into larger masses or flocules that eventually settle out of suspension

Flux : Movement and transport of mass of any water quality constituent over a given period of time. Units of mass flux are mass per unit time.

Flow Rate : The rate at which a fluid escapes from a unit area. Such measurements are made of liquid waste, effluent, and surface water movement.

Food Chain : Dependence of a series of organisms, one upon the other, for food. The chain begins with plants and ends with the largest carnivores

Forcing functions : External empirical formulation used to provide input describing a number of processes. Typical forcing functions include parameters such as temperature, point and tributary sources, solar radiation, and waste loads and flow.

Fresh Water : Water that generally contains less than 1,000 mg/L of dissolved solids.

Gradient : The rate of decrease (or increase) of one quantity with respect to another; for example, the rate of decrease of temperature with depth in a lake.

Groundwater : Phreatic water or subsurface water in the zone of saturation. Groundwater inflow describes the rate and amount of movement of water from a saturated formation.

Geochemical : Refers to chemical reactions related to earth materials such as soil, rocks, and water.

Grab Sample : A single sample collected at a particular time and place that represents the composition of the media only at that time and place.

Heterotrophic : Pertaining to organisms that are dependent on organic material for food.

Heavy Metals : Metallic elements with high atomic weights that can damage living things at low concentrations and tend to accumulate in the food chain, e.g., mercury, chromium, cadmium, arsenic, and lead.

Hydrodynamics : The study of the motion of, and the forces acting on, fluids

Hydrodynamic model : Mathematical formulation used in describing circulation, transport, and deposition processes in receiving water.

Hydrograph : A graph showing variation of in stage (depth) or discharge of water in a stream over a period of time.

Hydrology : The science that deals with the properties of the waters of the earth, their distribution on the surface and underground, and the circulation cycles involving evaporation, precipitation, flow, etc.

Hydrolysis : Reactions that occur between chemicals and water molecules resulting in the cleaving of a molecular bond and the formation of new bonds with components of the water molecule.

Hydraulic Radius : The right cross-sectional area of a stream of water divided by the length of that part of its periphery in contact with its containing conduit; the ratio of area to wetted perimeter. Also called hydraulic mean depth.

Initial conditions : A state of a system prior to an introduction of an induced stimulus. Describe conditions at the start up of system simulations.

Infiltration : 1) The penetration of water through the ground surface into sub-surface soil or the penetration of water from the soil into sewer or other pipes through defective joints, connections, or manhole walls. 2) The technique of applying large volumes of wastewater to land to penetrate the surface and percolate through the underlying soil. See Percolation.

Influent : Water, wastewater, or other liquid flowing into a reservoir, basin, or treatment plant.

Interstitial water : Water contained in the interstices, which are the pore spaces or voids in soils and rocks.

In situ : In place; in situ measurements consist of measurement of component or processes in a full scale system or a field rather than in a laboratory.

Kinetic processes : Description of the rate and mode of change in the transformation or degradation of a substance in an ecosystem.

Lagoon : 1) A shallow pond where sunlight, bacterial action, and oxygen work to purify wastewater; also used for storage of wastewater or spent nuclear fuel rods. 2) Shallow body of water, often separated from the sea by coral reefs or sandbars.

Load allocation (LA) : The portion of a receiving water's total maximum daily load that is attributed either to one of its existing or future nonpoint sources of pollution or to natural background sources.

Loading, Load, Loading rate : The total amount of material (pollutants) entering the system from one or multiple sources; measured as a rate in weight per unit time.

Longitudinal dispersion : The spreading of chemical or biological constituents, including pollutants, downstream from a point source at varying velocities due to the differential instream flow characteristics.

Littoral Zone : 1) That portion of a body of fresh water extending from the shoreline lakeward to the limit of occupancy of rooted plants. 2) The strip of land along the shoreline between the high and low water levels.

Mass balance : An equation that accounts for the flux of mass going into a defined area and the flux of mass leaving the defined area. The flux in must equal the flux out.

Mathematical model : A system of mathematical expressions that describe the spatial and temporal distribution of water quality constituents resulting from fluid transport and the one, or more, individual processes and interactions within some prototype aquatic ecosystem. A mathematical water quality model is used as the basis for waste load allocation evaluations.

Mineralization : The process by which elements combined in organic form in living or dead organisms are eventually reconverted into inorganic forms to be made available for a fresh cycle of plant growth. The mineralization of organic compounds occurs through combustion and through metabolism by living animals. Microorganisms are ubiquitous, possess extremely high growth rates and have the ability to degrade all naturally occurring organic compounds.

Mesotrophic : Reservoirs and lakes which contain moderate quantities of nutrients and are moderately productive in terms of aquatic animal and plant life.

Metabolites : Any substances produced by biological processes, such as those from pesticides.

Modeling : An investigative technique using a mathematical or physical representation of a system or theory that accounts for all or some of its known properties. Models are often used to test the effect of changes of system components on the overall performance of the system.

Monitoring : 1) Periodic or continuous surveillance or testing to determine the level of compliance with statutory requirements and/or pollutant levels in various media or in humans, plants, and animals. 2) Used to track the presence, migration, or threat posed by contaminants at a site; may be used at a site between response actions or when no other response action is appropriate until information or site status changes.

Monte Carlo simulation : A stochastic modeling technique that involves the random selection of sets of input data for use in repetitive model runs. Probability distributions of receiving water quality concentrations are generated as the output of a Monte Carlo simulation.

Mouth : The exit or point of discharge of a stream into another stream or a lake, or the sea.

Natural waters : Flowing water within a physical system that has developed without human intervention, in which natural processes continue to take place.

Numerical model : Models that approximate a solution of governing partial differential equations which describe a natural process. The approximation uses a numerical discretization of the space and time components of the system or process.

Nitrification : The process whereby ammonia in wastewater is oxidized to nitrite and then to nitrate by bacterial or chemical reactions.

Non-Point Source : Diffuse pollution sources (i.e., without a single point of origin or not introduced into a receiving stream from a specific outlet). The pollutants are generally carried off the land by storm water. Common nonpoint sources are agriculture, forestry, urban, mining, construction, dams, channels, land disposal, saltwater intrusion, and city streets.

Nutrient : A primary element necessary for the growth of living organisms. Carbon dioxide, nitrogen, and phosphorus, for example, are required nutrients for phytoplankton growth.

Organic : Refers to volatile, combustible, and sometimes biodegradable chemical compounds containing carbon atoms (carbonaceous) bonded together and with other elements. The principal groups of organic substances found in wastewater are proteins, carbohydrates, and fats and oils.

Oil Spill : An accidental or intentional discharge of oil which reaches bodies of water. Can be controlled by chemical dispersion, combustion, mechanical containment, and/or adsorption. Spills from tanks and pipelines can also occur away from water bodies, contaminating the soil, getting into sewer systems and threatening underground water sources.

Organic Matter : Carbonaceous waste contained in plant or animal matter and originating from domestic or industrial sources.

Outfall : The place where effluent is discharged into receiving waters.

Oxidation : The chemical union of oxygen with metals or organic compounds accompanied by a removal of hydrogen or another atom. It is an important factor for soil formation and permits the release of energy from cellular fuels.

Oxygen demand : Measure of the dissolved oxygen used by a system (microorganisms) in the oxidation of organic matter. See also biochemical oxygen demand.

Oxygen Use Rate : Rate of oxygen consumption due to biological and chemical action (used to determine respiration rate when the chemical oxygen demand is negligible).

Partition coefficients : Chemicals in solution are partitioned into dissolved and particulate adsorbed phase based on their corresponding sediment to water partitioning coefficient.

Pathogens : Microorganisms that can cause disease in other organisms or in humans, animals and plants (e.g., bacteria, viruses, or parasites) found in sewage, in runoff from farms or rural areas populated with domestic and wild animals, and in water used for swimming. Fish and shellfish contaminated by pathogens, or the contaminated water itself, can cause serious illness.

Percolation : 1) Movement under hydrostatic pressure of water, through the interstices of rocks or soils, downward to groundwater, except movement through large openings such as solution channels. 2) Slow seepage of water through a filter.

Physical and Chemical Treatment : Processes generally used in large-scale wastewater treatment facilities. Physical processes may include air-stripping or filtration. Chemical treatment includes coagulation, chlorination, or ozonation. The term can also refer to treatment of toxic materials in surface and groundwaters, oil spills, and some methods of dealing with hazardous materials on or in the ground.

Point source : Pollutant loads discharged at a specific location from pipes, outfalls, and conveyance channels from either municipal wastewater treatment plants or industrial waste treatment facilities. Point sources can also include pollutant loads contributed by tributaries to the main receiving water stream or river.

Pollutant : A contaminant in a concentration or amount that adversely alters the physical, chemical, or biological properties of a natural environment. The term include pathogens, toxic metals, carcinogens, oxygen demanding substances, or other harmful substances. Examples of pollutant sources include dredged spoil, solid waste, incinerator residue, sewage, garbage, sewage sludge, munitions, chemical waste, biological material, radioactive materials, heat, wrecked or discharged equipment, sediment, cellar dirt, hydrocarbons, oil, and municipal, industrial, and agricultural waste discharged into surface water or groundwater.

Pretreatment : The treatment of wastewater to remove or reduce contaminants prior to discharge into another treatment system or a receiving water.

Priority pollutant : Substances listed by the U.S. EPA under the Federal Clean Water Act as harmful substances and having priority for regulatory controls. The list includes metals (13), inorganic compounds (2), and a broad range of naturally occurring or artificial organic compounds (111).

Quality : A term to describe the composite chemical, physical, and biological characteristics of a water with respect to it's suitability for a particular use.

Raw sewage : Untreated municipal sewage.

Reaeration : The absorption of oxygen into water under conditions of oxygen deficiency.

Receiving waters : Creeks, streams, rivers, lakes, estuaries, groundwater formations, or other bodies of water into which surface water and/or treated or untreated waste are discharged, either naturally or in man made systems.

Residence time : Length of time that a pollutant remains within a section of a stream or river. The residence time is determined by the streamflow and the volume of the river reach or the average stream velocity and the length of the river reach.

Reverse Osmosis (RO) : A treatment process used in water systems by adding pressure to force water through a semi-permeable membrane, but containing contaminants. Reverse osmosis removes most drinking water contaminants. Also used in wastewater treatment. Largescale reverse osmosis plants are being developed.

River Basin : The land area drained by a river and its tributaries.

Runoff : That part of precipitation that flows to surface streams. Direct or overland runoff is that portion of rainfall which is not absorbed by soil, evaporated, or transpired by plants, but finds its way into streams as surface flow. That portion which is absorbed by soil and later discharged to surface streams is groundwater runoff.

Salinity : A measure of the amount of salt in water.

Sediment (SED) : 1) Material borne and deposited by water. 2) Soil, sand, and minerals washed from land into water, usually after rain. They pile up in reservoirs, rivers and harbors, destroying fish and wildlife habitat, and clouding the water so that sunlight cannot reach aquatic plants. Careless farming, mining, and building activities will expose sediment materials, allowing them to wash off the land after rainfall.

Sediment Yield : The quantity of sediment arriving at a specific location.

Sedimentation : 1) Letting solids settle out of wastewater by gravity during treatment. 2) Solids naturally settling out of slow water in rivers, streams and other water bodies.

Sensitivity : The slope of the analytical curve, i.e., functional relationship between emission intensity and concentration.

Settleable Solids : Material heavy enough to sink to the bottom of a wastewater treatment tank.

Settling Tank : A holding area for wastewater, where heavier particles sink to the bottom for removal and disposal.

Surface Water (SW) : All water naturally open to the atmosphere (rivers, lakes, reservoirs, ponds, streams, impoundments, seas, estuaries, etc.) and all springs, wells, or other collectors directly influenced by surface water.

Stratification (of water body) : Formation of water layers each with specific physical, chemical, and biological characteristics. As the density of water decreases due to surface heating, a stable situation develops with lighter water overlaying heavier and denser water.

Suspended Load : Sediment particles maintained in the water column by turbulence and carried with the flow of water.

Suspended Solids : Small particles of solid pollutants that float on the surface of, or are suspended in, sewage or other liquids. They resist removal by conventional means.

Swamp : A type of wetland dominated by woody vegetation but without appreciable peat deposits. Swamps may be fresh or salt water and tidal or non-tidal.

Tidal Marsh : Low, flat marshlands traversed by channels and tidal hollows, subject to tidal inundation; normally, the only vegetation present is salt tolerant bushes and grasses.

Topography : The physical features of a surface area including relative elevations and the position of natural and man-made features.

Toxic substances : Those chemical substances, such as pesticides, plastics, heavy metals, detergent, solvent, or any other material that are poisonous, carcinogenic, or otherwise directly harmful to human health and the environment.

Toxic Pollutants : Materials that cause death, disease, or birth defects in organisms that ingest or absorb them. The quantities and exposures necessary to cause these effects can vary widely.

Wastewater : The spent or used water from a home, community, farm, or industry that contains dissolved or suspended matter.

Water Pollution : The presence in water of enough harmful or objectionable material to damage the water's quality.

Watershed : The land area that drains into a stream; the watershed for a major river may encompass a number of smaller watersheds that ultimately combine at a common delivery point.

Zero Order Reaction : A chemical reaction in which an increase (or decrease) in reactant concentration results in no change in the rate of reaction (as long as some reactant is present).

❑❑❑

Index

B

C

D

E

F

G

H

I

J

K

L

M

N

O

P

Q

T

U

V

W

X

Y

Z

❑❑❑

FSC
www.fsc.org
MIX
Papier | Fördert gute Waldnutzung
FSC® C083411

Zeitfracht Medien GmbH
Ferdinand-Jühlke-Straße 7
99095 Erfurt, Deutschland
produktsicherheit@kolibri360.de